Biological Control of Plant Pathogens, Weeds and Phytoparasitic Nematodes

Biological Control of Plant Pathogens, Weeds and Phytoparasitic Nematodes

M. Reddi Kumar
M. John Sudheer
B. Chenchu Reddy
B. Srinivasulu
Acharya N.G. Ranga Agricultural University.
Hyderabad.

BSP **BS Publications**
A unit of **BSP Books Pvt., Ltd.**
4-4-309/316, Giriraj Lane, Sultan Bazar,
Hyderabad - 500 095
Phone : 040 - 23445605, 23445688

Published by :

BSP BS Publications

A unit of **BSP Books Pvt., Ltd.**

4-4-309/316, Giriraj Lane, Sultan Bazar,
Hyderabad - 500 095
Phone : 040 - 23445605, 23445688
e-mail : info@bspbooks.net

ISBN : 978-93-52300-64-8 (HB)

Acharya N.G.Ranga Agricultural University
Rajendranagar, Hyderabad-500 030.
Phone : Off : 010-24015035
Res : 040-27425305
Grams : 'Agrivarsity'
Fax : 040-24015031
E-mail : angrau_vc@yahoo.com

S. Raghava Reddy
M.sc. (Ag.), Ph.D.
Vice-Chancellor

Message

Of late, the sharp increase in the use of chemical pesticides has resulted in the development of fungicidal and insecticidal resistance in key pest species. The biopesticides developed based on microbials such as fungi, bacteria and nematodes were studied mostly to develop alternatives to chemicals. A tremendous break-through has been witnessed in the research efforts on biological control of plant diseases especially by using species of *Trichoderma, Gliocladium* and *Pseudomonas* in India. Among bacteria, the fluorescent pseudomonads form the major group found surviving in rhizosphere of all crops. The Plant Growth Promoting Rhizobacteria (PGPR), known for their high biocontrol efficacy, which are reported to control several soil, seed and air-borne crop diseases caused by different plant pathogens.

The work on development of antagonists as myco*herbicides has gaining* mo*mentum because o*f the increased awareness of threat from the weeds. In India, considerable work has been done with *Alternaria eichhorniae* for managing water hyacinth. Biological control of plant parasitic nematodes has become most significant and intensive areas of research because of environmental concerns and non-availability of chemicals. Exploitation of natural ellemies in the management of plant parasitic nematodes seems to be practically demanding and a relevant approach in the present context. Of the microorganisms, *fungi hold an important* position and some of them have shown great potential as biocontrol agents.

Pertain to this, the book entitled **"Biological Control of Plant Pathogens, Weeds and Phytoparasitic Nematodes"** authored by Dr.M.Reddi Kumar, Asst. Professor, Dept. of Plant Pathology, S.V. Agricultural College, Tirupati, Dr. M.John Sudheer, Senior Scientist, Agriculture Research Station, Kadiri, Dr.B.Chenchu Reddy, Director, International Programmes, Hyderabad and Dr.B.Srinivasulu, Principal Scientist, Ambajipeta provides an insight into different aspects of biological control such as isolation and selection of potential biocontrol agents, genetic improvement, several specific techniques and procedures in formulation technology etc. have been described.

I congratulate the authors for their studious efforts in bringing out the book on a specified area of vast expanse. I hope this book would serve as a reference material for all those who are dwelling out biological control studies.

6/2/08
(P. Raghava Reddy)

Dr. M. Sudarshan Reddy
M.sc. (Ag.), Ph.D.
Dean of Agriculture

Acharya N.G.Ranga Agricultural University
Rajendranagar, Hyderabad-500 030.
Phone : Off : 010-24015035
Res : 040-27425305
Grams : 'Agrivarsity'
Fax : 040-24015031
E-mail : angrau_vc@yahoo.com

Foreword

Over the past one hundred years, research has repeatedly demonstrated that phytogenetically diverse microorganisms can act as natural antagonists of various plant pathogens. Biological control of plant pathogens continues to inspire research and development in many fields. Plant pathogens are one class of target of biological control, which also is designed to limit other pests such as insects, parasitic nematodes and weeds. Some of the microbial taxa that have been successfully commercialized and marketed as biopesticides, include bacteria belonging to the genera *Agrobacterium, Bacillus, Pseudomonas* and *Streptomyces* and fungi belonging to the genera *Ampelomyces, Candida, Coniothyrium* and *Trichoderma* after intensive screening.

Advanced molecular techniques are now being used to characterize the genetic diversity, abundance and activities of microbes that live in and around plants, including those that significantly impact plant health. Successful application of biological control requires more knowledge-intensive management and emphasis should be laid on practical aspects such as mass production and formulation to make new biocontrol products stable, safer and cost-effective.

In this connection , the book entitled **"Biological Control of Plant Pathogens, Weeds and Phytoparasitic Nematodes",** authored by Dr. M. Reddi Kumar, Asst. Professor, Dept. of Plant Pathology, S.V. Agrl. College, Tirupati, Dr. B. Srinivasulu, Principal Scientist, Ambajipeta and Dr. B. Chenchu Reddy, Associate Dean, S.V. Agrl. College, Tirupati and Dr. M. John Sudheer, Senior Scientist, ARS, Kadiri gives the basic and applied information in the field of biological control of plant diseases. The information has been provided in a lucid and vivid manner to understand the basic concepts of the subject, essential for both undergraduate and postgraduate students.

In the present world of competition, students can excel only if they are competitive and informative. I hope that this book would provide useful information to students, research scholars and teachers as well. I appreciate the efforts made by the authors in bringing out the book on biological control.

(M. Sudarshan Reddy)

Preface

The sharp increase in the use of chemical pesticides in India in recent years has resulted in severe implications in the development of fungicidal and insecticidal resistance in key pest species, pesticide residues in food chain, degradation in the quality of ecosystem and human health. Microbials such as viruses, bacteria, fungi, protozoa, nematodes and plant products are the major biopesticides that were studied mostly to develop alternatives to chemicals. In India, biopesticide science is not a new tool and is as old as human civilization back to prehistoric days. Though biopesticides cover only about 1% of the total plant protection products globally, their number and growth rate have been showing an increasing trend in the past two decades.

The last decade has witnessed a tremendous breakthrough in the research efforts on biological control of plant diseases in India especially by using species of *Trichoderma*, *Gliocladium*, *Bacillus* and *Pseudomonas*. Several economically important plant diseases of a variety of crops could be effectively managed either by soil application or seed treatment. The fungal antagonists that have shown considerable potential in controlling several plant pathogenic fungi, particularly soil-borne ones and commercial formulations available have been mentioned. In this book, several specific techniques and procedures in formulation technology, critical for the practical and effective implementation of biocontrol, have been described briefly.

Fluorescent pseudomonads form the major bacterial group found surviving in rhizosphere of all field crops. They are also known as "Plant Growth Promoting Rhizobacteria"(PGPR) as they promote plant growth by secreting auxins, gibberellins and cytokinins. *Pseudomonas fluorescens*, *P. putida* and *P. cepacia* are the important fluorescent pseudomonad species which are known as highly effective biocontrol agents. They have been reported to control several soil, seed and air-borne crop diseases caused by fungi, bacteria and viruses. Emphasis is being laid on selection of effective strains of bacteria, mechanisms of action and suitable delivery systems.

The menace of aquatic weeds is reaching alarming proportions, especially in the tropical and sub-tropical zones where warm weather supports their profuse growth. Of the seven nuisance aquatic weeds of the world, three weeds viz; water hyacinth, hydrilla and Eurasian watermilfoil have remained the focus of attention. Though a few successes have been achieved on the biocontrol of aquatic weeds, control of submerged aquatic weeds with fungal pathogens on a large scale, is yet to commence. One of the objectives of the book is to aquaint the readers with aquatic weeds of worldwide importance and the status of research in controlling three aquatic weeds with fungal pathogens.

Biological suppression of plant parasitic nematodes with fungal and bacterial antagonists is gaining increasing importance as a result of realization that many environmental and health hazards are linked with the use of chemicals. Standardization of methods for effective utilization of biocontrol agents is very important for evolving ecologically sound integrated nematode management strategies. The role of most important bacterial and fungal antagonists in biological suppression of plant parasitic nematodes on fruits, vegetables and plantation crops has been discussed.

The research and development leading to commercial scale production of antagonistic organisms for biological control of plant diseases has been quite intense in the recent past. Though lot of reports are available on identification of strains of biological control agents, their mass production, formulation and field application, the books available dealing with the above aspects are scarce. Hence, the present work was started with the intention of elucidating concepts involved in biological control, techniques of isolating effective strains of bioagents, fermentation technology and delivery systems, with the hope to help the students in understanding the subject and the latest developments that have been witnessed. The information presented in this book has been collected from various review articles, technical bulletins and online publications.

We are grateful to Dr. P. Sudarshan Reddy, Dean of Agriculture, ANGRAU, Hyderabad for his constant support and Dr. K. Veeranjineyulu, University Librarian for his encouragement in accomplishing this work. We hope this book would be useful to graduate, postgraduate students and research scholars who are pursuing biological control studies.

–Authors

Contents

CHAPTER 4

Molecular Characterization of Biocontrol Agents

Chapter 5

Morphological Charateristics of *Trichoderma* Species

Chapter 6

Mass Production and Formulation of *Trichoderma spp.*

Chapter 7

Formulation and Shelf-Life of *Trichoderma spp.*

Chapter 8

Shelf-Life and Bio-efficacy of Biocontrol Formulations

Chapter 9

Delivery of Biocontrol Agents

CHAPTER 12

Phytoalexins in Plant Disease Control

CHAPTER 13

Mycorrhizae - A Tool in Biological Control

Chapter 14

Plant Growth Promoting Rhizobacteria

CHAPTER 15

Biological Suppression of Weeds with Pathogens

CHAPTER - 16

Biological Control of Plant Parasitic Nematodes

Introduction

Biological control assumes major importance in sustainable agriculture, especially where soil-borne plant disease problems are serious. The last decade has witnessed a tremendous breakthrough in the research efforts on biological control of diseases in India, especially by using *Trichoderma, Gliocladium, Pseudomonas* and *Bacillus*. The development of biological control technologies is a newly emerging field in Plant Pathology compared to the advances made with these systems to reduce damage to economic crops caused by insects and weeds. Several economically important plant diseases of a variety of crops could be effectively managed either by soil application or by seed treatment with fungal and bacterial bioagents.

It is becoming increasingly apparent throughout the world that efficient and productive agriculture is practiced at great cost to the environment. There are increasing reports of loss of pesticides from the market due to the resistance development in the pest population, groundwater and food stuff pollution. Many fungicides used for plant disease control possess potential oncogenic risks. Fungal pathogens are important biotic factors of various crops. In India, a great deal of work has been done on control of soil-borne pathogens like *Sclerotium rolfsii, Rhizoctonia, Pythium aphanidermatum, Sclerotinia sclerotiorum, Phytophthora spp.* and *Fusarium spp.* using *Trichoderma and Gliocladium virens.*

There is need to continuously look for antagonists that are able to colonize plant surface and survive in competitive environment of the soil. Microorganisms show many traits which may characterize them particularly adept at either the colonization phase or subsequent phase of competition. Such species have a high reproductive capacity apart from possessing the ability to produce effective antimicrobial compounds like antibiotics, siderophores, volatile compounds etc. Antagonistic bacteria show promising ability in biocontrol of plant pathogens. However, their successful use depends on judicious management and can form an effective component of integrated disease management. Managing highly susceptible varieties in hot spot areas with mainly antagonistic bacteria will be difficult. However, combining modest level of plant resistance with such antagonists could provide valuable results.

Biocontrol of foliar diseases has receive little attention in India, compared to soil-borne diseases. This is in contrast with the availability of a number of formulated products of antagonists for soil-borne disease control. In India, having a tropical climate and inadequate storage facilities, a huge amount of agricultural produce is lost in storage and transit due to post-harvest pathogens. Very few attempts were made in India to control storage pathogens using bioagents, although a couple of commercial formulations of antagonists are already available in some countries for post-harvest applications. The major stumbling block in the success of biological control technologies is though it has been possible to isolate and demonstrate the efficacy of antagonists in laboratory, greenhouse and experimental field conditions, but rarely in farmer's field. The major aspects for consideration of successful biological control technologies include the establishment of production, formulation and delivery systems for bioagents. In addition, these aspects must be implemented to be compatible with industrial and commercial development methods, and field applications. Liquid fermentation technology, which has many advantages over solid state fermentation in production of bioagents, needs to be standardized. An important consideration during the entire research and development process should be careful monitoring for quality control of the product. Research priorities must include the following broad areas before several *Trichoderma, Gliocladium* and other bioagent strains become available commercially for biocontrol.

- Development of computerized models to understand the effect of microenvironment on the survival and proliferation of antagonists in soil and plant rhizosphere.
- Genetic manipulation to enhance the ability of the antagonists to control several diseases adapt to various environmental conditions and tolerate pesticides.
- Understanding the mechanism of biocontrol at molecular and biochemical level.
- Development of fermentation technology and delivery systems.

In recent years, the menace posed by aquatic weeds has attained alarming proportions, especially in the tropics and sub-tropics. There is persistence of the nuisance caused by the weed to human affairs including agriculture, notwithstanding the considerable success achieved through the imported arthropods for the control of water hyacinth. The work on mycoherbicides has been directed towards the weeds over past three decades, since water hyacinth has achieved the status of the number-one aquatic weed in India. In Florida, a successful mycoherbicide based on *Cercospora rodmanii* has been developed for control of water hyacinth.

The task of bringing out commercial product has been highly difficult right from its discovery in early 1970's to its formulation. Considerable work has been done in India with *Alternaria eichhorniae* for managing water hyacinth. Several fungal pathogens have been isolated, of which *Myrothecium roridum* and *Verticillium nigricans* hold good promise.

Among the terrestrial weed problems, *Parthenium hysterophorus* is the most obnoxious weed in our country posing dangers to humans, animals and environment as a whole. The work on development of its antagonists as mycoherbicides has received a boost in recent years because of the increased awareness of the threat from the weed. Leaf feeding *Zygogramma bicolorata*, the Mexican beetle, introduced in 1980s has not been able to completely suppress the weed because of enormous seed production habit. Biological control with pathogens in addition to the insect appears to be the best option to check further spread of the weed. Rust pathogen, *Puccinia abrupta* var. *partheniicola* used as a classical bioagent in Australia, has shown some dramatic results in the suppression of the weed.

Like many other countries, plant parasitic nematodes are significant impediments in agricultural productivity in India. Root-knot, cyst forming, reniform, ear-cockle, burrowing and lesion nematodes besides several ectoparasitic nematodes are distributed throughout the country. Though adequate information from different regions of the country is not yet available or collected, some localized but intense nematode problems like cereal cyst nematode in Rajasthan on wheat and barley, potato cyst nematode on potato in the Nilgiri hills of Tamil Nadu, burrowing nematode on banana in Karnataka and Kerala and coconut, areca nut and black pepper in Kerala and root-knot nematode on tobacco and vegetables are of major concern.

Of late, biological control of plant parasitic nematodes has become most significant and intensive area of research because of environmental concerns and non-availability of effective nematicides. Many natural enemies such as fungi, bacteria, viruses, predatory nematodes, insects, mites and some invertebrates have been found to parasitize or prey upon the nematodes. Exploitation of these enemies for practical use in the management of plant parasitic nematodes seems to be practically demanding and is a relevant approach in the present context. Of the microorganisms, fungi hold an important position and some of them have shown great potential as biocontrol agents. Some 70 genera and 180 species of fungi have been found associated with nematodes but only few of them are successful biocontrol agents. The fungi involved in biocontrol of nematodes are endoparasitic, predacious and opportunistic fungi. Assessment of potential of endoparasitic and predacious fungi in biocontrol has not received much attention except *Hirustella rhossiliensis* and *Arthrobotrys oligospora.*

Compared to other two groups, opportunistic fungi are gaining more importance in biocontrol programmes. These fungi are capable of colonizing nematode reproductive structures and deleteriously affecting them. Nematodes belonging to *Meloidogyne, Heterodera and Globodera* at sedentary stage of their life cycle are vulnerable to attack by the fungi either within the host or in soil. They also attack nematode eggs. *Verticillium chlamydosporium, Paecilomyces lilacinus, Exophiala* sp. and *Cylindrocarpon* sp. are the fungi which are usually encountered. The fungi, *V. chlamydosporium* and *P. lilacinus* have been shown to be capable of preventing egg hatching and colonizing eggs by hyphal penetration. Enzymatic disruption and physiological disturbance brought about by biosynthesis of diffusible toxic metabolites are thought to be the main types of destructive activity. *Gliocladium, Trichoderma and Acremonium* are the other opportunistic fungi, which are also found to affect nematode multiplication but the mechanism of suppression of nematode population is lacking.

An obligate nematode bacterium, *Pasteuria penetrans* is known to infect root-knot and cyst nematodes and has considerable potential as a biocontrol agent. The bacterium commences infection when endospores attach to the cuticle of second stage juveniles in soil. The germ tube of endospore penetrates the cuticle of the nematode, enter the body and the spores fill entire body, preventing egg formation. *Pasteuria* spp. have been reported from over 135 species of nematodes belonging to 50 genera. At present, only three species viz., *P. penetrans, P. thornei, P. nishizawae* are found to be associated with nematodes. Even though, *P. penetrans* appears to have considerable potential as biocontrol agent, commercial exploitation is prevented by lack of large scale production methods. An *in vivo / in vitro* method of production is needed where the active growth of bacterium can be maintained without its nematode host.

For mass production of nematode biocontrol agents, necessary fermentation technologies are needed for commercialization of biological control agents. Finding a suitable medium using inexpensive and readily available materials is the first step in commercial fermentation. Little is known about the nutrient requirement of nematophagous fungi and bacteria. The nematophagous fungi viz. *V. chlamydosporium* and *P. lilacinus* are facultative parasites which can grow on a wide range of culture media. The fungi colonize solid substrates like bran, straw, wheat or rice kernels and a wide variety of standard laboratory media but grow readily in submerged culture in liquid media.

Very little has been studied about the formulation of nematode biocontrol agents. Granular formulation is considered to be most suitable for microorganisms that are to be applied to soil. Different formulations such as sodium alginate prills, diatomaceous earth granules, lignite-stillage granules, alginate-clay pellets, talc and kaolin-based products were proved suitable for carrying nematode biocontrol agents. For the biocontrol of root-knot nematode, *P. lilacinus* is commercially available as '**Biocon**' in the Philippines. *Arthrobotrys robusta* is commercially formulated as '**Royal 300**' while '**Royal 350**' is a similar product containing *A. superba*. Two strains of predatory fungus, *Arthrobotrys* are available as pesticidal agent. **Nematofagin– BL** in Russia are effective against several nematodes in various crops.

Impact of epidemiological factors on the various biocontrol agents in the field is an important aspect of biocontrol strategy. Effectiveness of antagonist and its ability in an ecosystem, host exudates, chemicals and the environment are important factors to decide the fate of biocontrol of plant pathogens. Rhizosphere is preferred by the biocontrol agents over phyllosphere for their survival. It is, therefore, of much importance to investigate more on the epidemiological aspects of diseases and biocontrol agents to achieve greater success in this eco-friendly and relatively cheaper technology.

Thus, the utilization of biocontrol agents in a well-orchestrated IDM strategy encompassing moderately resistant varieties, time of planting, fertilizer and organic manures and with minimal dependence on pesticides as and when required, will pave the way for a pollution-free environment.

Chapter - 1

Isolation and screening of Bioagents of Plant Pathogens

1.0 Isolation of Antagonistic Fungi

There are several fungi which can be used as biocontrol agents for control of plant diseases viz; *Trichoderma, Gliocladium, Aspergillus, Penicillium, Myrothecium* sp. etc., (Fig. 1.2 & 1.3). Among these, *Trichoderma* and *Gliocladium* are ubiquitous in nature i.e., its presence everywhere. They are commonly present in a wide variety of habitats such as decaying organic matter, manure heaps, crop residues, rhizosphere and non-rhizosphere soils etc.

1.1 Collection of Soil Samples for Isolation of Bioagents

For isolation of bioagents, soil samples are to be collected from the field where disease occurrence is low but pathogen is known to be present. Then chances of isolating an efficient strain will be more in areas where the pathogen is introduced but not established and also in areas of monoculture of crops where disease intensity has decreased after a few years with a susceptible crop. In cases where the particulars of previous crop and disease occurrence is not known, biologically active soils that contain diverse population of microorganisms could be sampled. Soils of special interest should also be taken into account in some cases. Soil should be collected at a depth of 15 cm in the upper surface including rhizosphere and rhizoplane samples. Phylloplane microflora can be isolated from healthy foliar parts such as flowers, fruits, and leaves of diseased plants. After collection of soil samples, isolation part should be completed as soon as possible. (Fig. 1.1).

1.2 Isolation of *Trichoderma* and *Gliocladium* spp.

Primarily, there are three different areas that can be tapped for isolation of bioagents such as rhizosphere, rhizoplane and phylloplane.

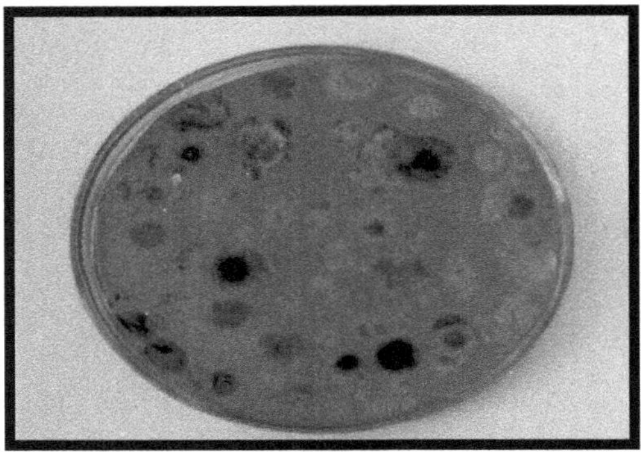

Fig. 1.1 Colonies of microflora isolated from groundnut rhizosphere.

Pure Cultures of soil microflora isolated from rhizospheric soil

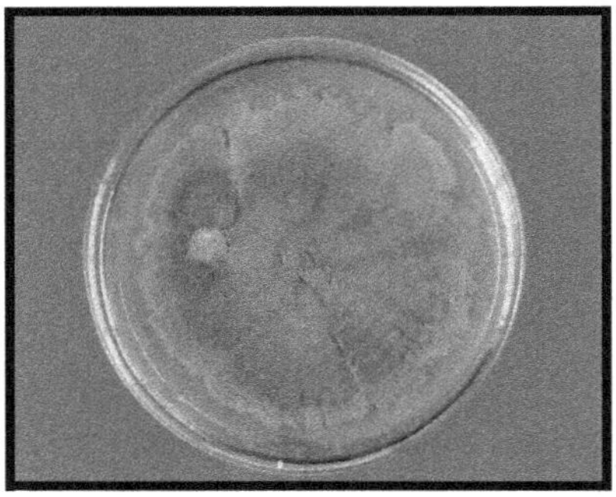

***Penicillium* sp.**

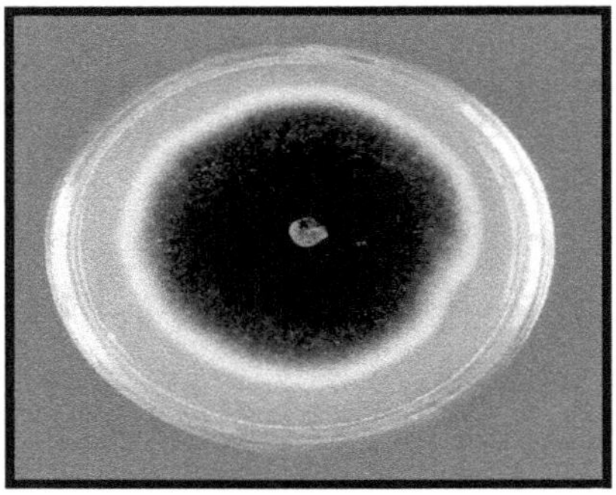

Aspergillus niger

Fig. 1.2 Pure cultures of *Penicillium sp.* and *Aspergillus niger.*

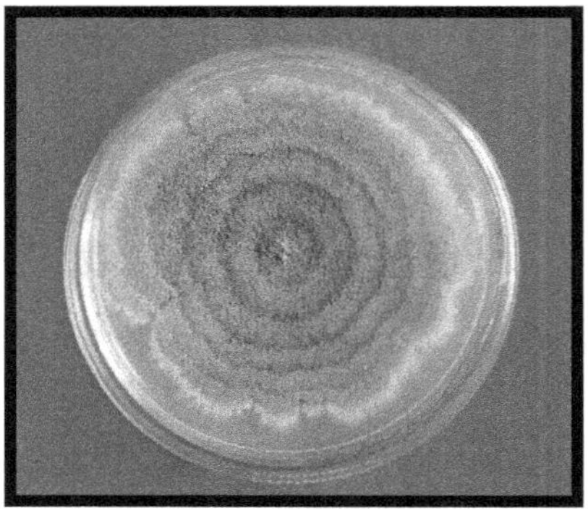

Aspergillus flavus

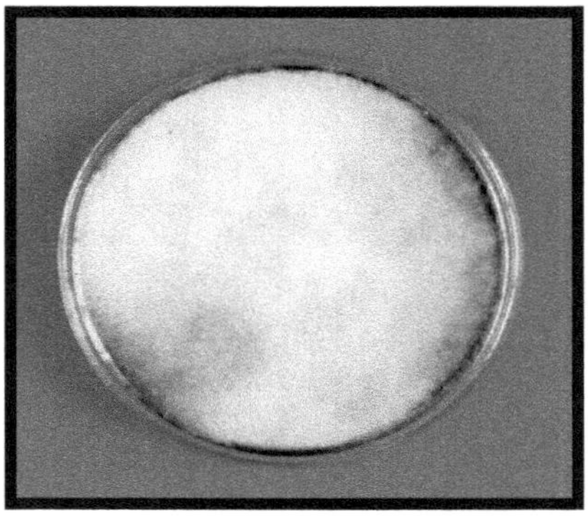

Rhizopus stolonifer

Fig. 1.3 Pure cultures of *Aspergillus flavus* and *Rhizopus stolonifer.*

1.2.1 Isolation from Rhizosphere

After collection of rhizospheric soil samples, separate the soil adhered to roots of the plant with the help of a brush and collect it in a Petri plate. Take 10 gm of soil sample in a conical flask containing 100 ml sterile distilled water and keep it for shaking for about 15 minutes on a magnetic stirrer. From the dilution blank, prepare serial dilutions, of 10^{-2} to 10^{-6}. Pipette out 1 ml of each dilution of 10^{-4} to 10^{-6} to sterile Petri plates. Pour about 20 ml of *Trichoderma* selective medium and agitate the plate gently so as to ensure uniform distribution of propagules. The plates are to be incubated at 20°C for 5-7 days. Transfer the isolated colonies into PDA slants and prepare microscopic slides for observation under microscope. (Fig. 1.4).

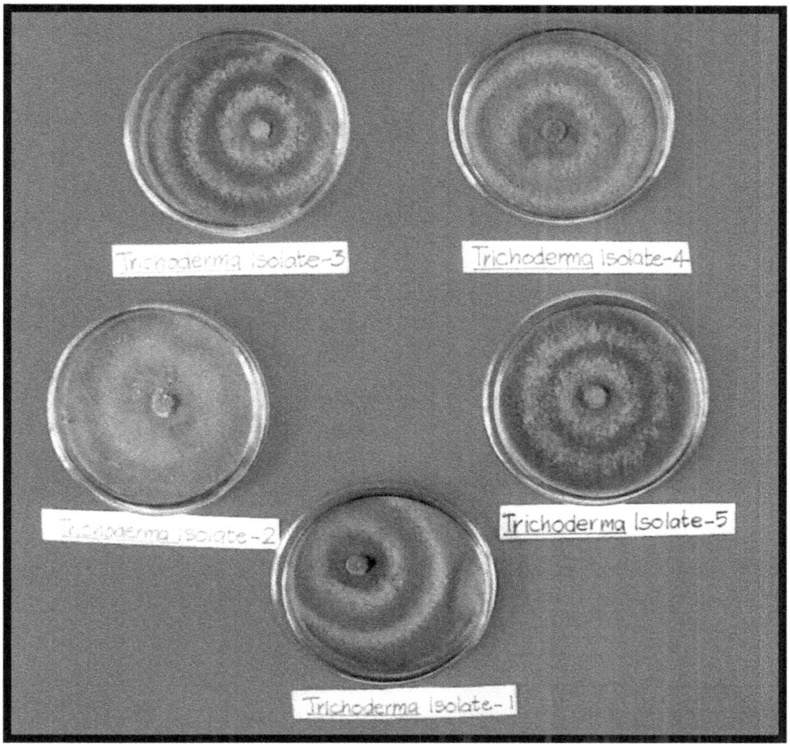

Fig. 1.4 Pure cultures of *Trichoderma* spp.

1.2.2 Isolation from Rhizoplane

For isolation of microorganisms associated with the surface of roots, collect the root pieces into a screw cap bottle containing 100 ml of sterile distilled water and shake it well. Then the root pieces are subjected to serial washings with sterile distilled water until clear root surface is exposed. Plate the washed roots on a specific medium and incubate the plates at 25°C for 5-7 days and observe for colony growth.

1.2.3 Isolation from Phylloplane

For obtaining colonies of antagonistic fungi from phylloplane, fresh and healthy leaves of all ages are to be collected in sterile polythene bags. From every leaf using sterile cork borer, transfer about 50 discs to 100 ml sterile water blank and stir for 20 minutes using magnetic stirrer. From this water blank, prepare serial dilutions of 10^{-1} and 10^{-2}. Transfer 1 ml aliquot from all dilutions to sterile Petri plates and pour *Trichoderma* specific agar medium and gently agitate the plates for uniform distribution. Incubate the plates at 25°C for 5-7 days and observe for growth of colonies of antagonistic fungi.

1.3 Isolation of Antagonistic Bacteria

Rhizosphere bacteria like fluorescent Pseudomonads, *Bacillus subtilis* and *Streptomyces* spp. have been successfully utilized as biocontrol agents for control of plant diseases which form an important alternative to the chemical control. Isolation of rhizospheric bacteria is easy as they are ubiquitous but selection of an efficient strain is difficult since only a few of them will be disease suppressors. Hence, selection of a potent strain of rhizospheric bacteria must be based on its **rhizosphere competence, antagonistic ability** and **root colonizing ability**. The nutrient-rich rhizosphere of plant roots is naturally colonized by many beneficial or pathogenic bacteria and fungi. Beneficial microorganisms improve uptake of nutrients and/or produce plant growth promoting substances. They also protect plant surfaces from colonization of pathogenic microorganisms through either direct competitive effects or production of antimicrobial compounds.

The antagonistic bacteria can be isolated from different regions, namely **(a) Rhizosphere (b) Foliage or Phyllosphere and (c) Endosphere or endophytic.**

1.3.1 Isolation from Rhizosphere

Isolation of antagonistic bacteria associated with roots of plants can be done by picking out the whole roots along with closely adhering soil from healthy plants growing in soil where disease is present or seen earlier. Place whole roots in a conical flask containing 100 ml sterile distilled water and constantly stir for about 30 minutes to dislodge the rhizosphere associated bacteria. From this dilution blank, prepare serial dilutions upto 10^{-5} and pipette out 1ml of sample from each dilution on different media viz., King's B medium, Soil extract agar, *Pseudomonas* agar, Ken Knight's agar medium etc. Incubate the plates at a temperature of 28 °C for about 48 hrs and observe for colony growth. As numerous colonies of bacteria will emerge on the surface of media, select discrete individual single colonies and transfer a loopful of bacterial culture onto slants for purification. Select distinct bacterial colonies based on color and morphology and screen the isolated bacteria for their ability to antagonize plant pathogens.

1.3.2 Isolation from Phyllosphere

Phyllosphere bacteria can be isolated from the surface of healthy leaf or from other leaf structures. For the purpose of isolation, collect healthy leaves from a diseased plant or plants growing in disease conducive soils.

Collect healthy, mature leaves from plants growing in disease conducive soils. To remove dust particles and loosely adhering unassociated microbes, the collected leaves are subjected to gentle wash with sterile water for few seconds and then the leaves are collected in a conical flask containing 100 ml sterile distilled water and vigorously shaken for 30 minutes. Make serial dilutions from the dilution blank and pipette out 1 ml of sample onto different media viz., King's B medium, Ken Knight agar medium, Tryptic soy agar medium (TSA), *Pseudomonas* agar etc. The plates are to be incubated at a temperature of 28°C for 48 hrs. For obtaining pure cultures, transfer a loopful of discrete colony onto a fresh TSA plate and screen the isolated bacteria for its antagonism against plant pathogens.

1.4 Screening and Selection of Potential Biocontrol Agents

1.4.1 *In vitro* Screening Methods for Antagonistic Fungi

Efficacy of test fungi for their ability to contain plant pathogenic fungi under *in vitro* can be studied by different ways such as,

(a) Dual culture plate technique.

(b) Antibiosis test for production of volatile compounds.

(c) Test for production of antibiotics in culture filtrate.

(d) Antibiosis test for production of diffusible or non-volatile compounds.

The overall ability of the test fungus in inhibiting the pathogen through competition for nutrients, production of inhibitory compounds and hyperparasitism can be studied in dual culture method, while antibiosis tests are used to determine the ability of the test fungus to produce volatile and non-volatile metabolites to inhibit the pathogenic fungi. The culture medium used for testing should be favourable for both antagonist and pathogen.

1.4.2 Dual Culture Method

It is one of important technique used to evaluate the antagonism by inoculating pathogen and opposing fungus. Six mm discs of pathogen and test fungus (antagonistic fungus) cut from 5-7 days old cultures are to be placed on potato dextrose agar (PDA) or malt agar plates (2%) in such a manner that they would lie opposite to each other at the corners of plate and incubated plates inoculated with either pathogen or test fungus alone in a similar position would serve as control. Three replications can be used in each case and the plates are to be incubated for 5-15 days. The radial growth of both pathogen and test fungus is to be measured in dual and monoculture plates after 24 hrs of incubation (Fig. 1.5). The percentage inhibition of pathogen/test fungus can be calculated by using these measurements. Based on observations taken in dual cultures, the type interaction occuring between pathogen and test fungus may be identified using the key consisting of following types of interactions.

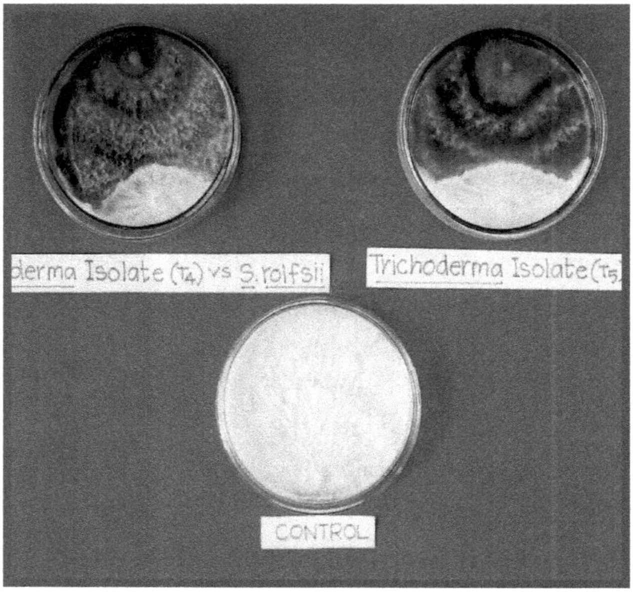

Fig. 1.5 *In vitro* efficacy of *Trichoderma* isolates on the mycelial growth of *Sclerotium rolfsii.*

(i) ***Overgrowth category (O) :*** Overgrowth of antagonistic fungus on the pathogen associated by the inhibition of the pathogen colony on or after contact.

(ii) ***Extreme inhibition of pathogen (E) :*** Extreme inhibition of pathogen at a distance of 5mm or more and little or no reciprocal inhibition of the opposing fungus.

(iii) ***Intermingling category (I) :*** Intermingling of colonies without macroscopic/microscopic interaction.

(iv) ***Mutual inhibition with pigmented band (MP) :*** Mutual inhibition of pathogen and antagonistic fungus with a pigmented band of mycelium at point of contact.

(v) ***Mutual inhibition with a clear zone (MC) :*** Mutual inhibition of pathogen and opposing fungus with a clear zone between the colonies.

In dual culture method, as the two organisms grow towards each other, production of antibiotic is indicated by reduced growth rate of the pathogen at a distance from the periphery of potential antagonist, whereas parasitism is indicated when the pathogen stops growing upon contact with the antagonist and its mycelium begins to lyse backwards and the antagonist may continue its growth over the colony of the test fungus. Other reactions that can occur in dual culture are that both organisms stop growing upon contact, with a small but clearly marked space between them.

Hyphal Interaction Studies

Hyphal interaction in dual culture can be studied by growing the fungi on a cellophane membrane in Petri dish with water-agar plus glucose (0.5%w/v). After 4 to 5 days, pieces of cellophane membrane from the contact zone are cut and prepared for microscopic observation. Different ways of parasitization of antagonist over the pathogen viz., penetration of mycelia, coiling, production of specialized absorptive organs, vacuolization etc., can be studied under magnification of 400x.

1.4.3 Antibiosis Test for Production of Non-volatile Metabolites

To know the production of diffusible or non-volatile compounds by the antagonistic fungus, cellophane paper method is followed. For this test, cellophane paper discs of 90 mm diameter having a thickness of 50 μm are cut and sterilized in an autoclave at 121°C for 15 min and placed carefully over sterilized PDA/malt agar medium (2%) and are taken into a Petri plate. From the fully grown plate of antagonistic fungus, six mm discs are cut from the periphery of fungus growth and placed at the centre of cellophane paper and incubate for about 2-5 days.

After incubation, remove the cellophane paper and the adhering fungus carefully. A six mm disc of the pathogen may be placed immediately on the medium at a central position previously occupied by the antagonistic fungus. The radial growth of the pathogen is recorded after every 24hrs and compare with the growth of pathogen in control plate. Based on this, the percentage inhibition of the pathogen can be calculated.

1.4.4 Test for Production of Antibiotics in Culture Filtrate

The test is similar to poisoned food technique, used for evaluating the efficacy of fungicides *in vitro*. In this method, the antagonistic fungus is grown in a suitable liquid medium either in stationary or shake culture. After sufficient growth of the fungus on the medium, the fungus mycelial fragments and other cells are removed by passing through filter paper and it is sterilized by sintered glass filter. The sterile culture filtrate without addition of sterile distilled water would be taken as cent percent concentration. The sterile filtrate in desired concentration (S, S/2, S/3) is added to sterile culture medium and plated. After solidification, the test pathogen is inoculated at the centre of the plate. Antagonistic fungus grown on PDA not amended with culture filtrate would serve as control. Reduction in colony growth in filtrate-amended medium compared to control indicates the secretion of antibiotic compounds into culture by antagonistic fungi.

1.4.5 Test for Production of Volatile Compounds

It is modified version of sealed Petri plate technique described by Dennis and Webster (1971b). For this purpose, two Petri dish bases containing 20 ml of solidified potato dextrose agar or malt agar medium are sealed together by an adhesive tape and six mm disc of antagonistic disc is inoculated at the centre of the medium allowed to grow for three, six and ten days in incubator at a temperature of $25 \pm 2\ ^{\circ}C$ and simultaneously allowing the medium present on the upper plate to absorb any volatile compounds produced during the period of exposure by the antagonistic fungus. After sufficient period of exposure, the upper plate containing the medium is carefully removed and inoculated with the pathogen. The radial growth of the pathogen is to be recorded after every 24 hrs and compared with the growth of the pathogen in control. Based on the observations recorded, percentage inhibition of the pathogen can be calculated.

1.5 Selection of Isolates

In most of the cases, selection of antagonistic isolates of *Trichoderma* under *in vitro* conditions is carried out just by means of dual culture technique. In this case, weightage is given only for the strain that produces maximum inhibition zone, which reflects the antibiotic production. However, in many cases the antibiotics produced by the antagonist are not found to be sufficient to act against the pathogens present in the soil as the antibiotics are adsorbed by the clay particles. Since other modes of action such as hyperparasitism, colonization behaviour, pathogen inhibition (PI), competitive saprophytic ability (CSA) and production of lytic enzymes are not given weightage in the selection of antagonists. The modified Antagonism Index concept can be used for the selection of best performing virulent antagonists.

1.5.1 Antagonism Index Concept

Antagonistic potential of *Trichoderma* was assessed against *Rhizoctonia solani, Macrophomina phaseolina, Sclerotium rolfsii, Fusarium udum* and *Macrophomina phaseolina* by using a new antagonism index concept following criteria viz., (PI), (CB), (IZ), (SOOP), (PL) and (CSA). The antagonism index of each strain can be assessed using the formula used by Nakkeeran et al., 2001.

$$\mathbf{AI = CSA \times CB \times PI \times PL \times SOOP \times IZ}$$

1.5.2 Competitive Saprophytic Ability (CSA)

To assess the competitive saprophytic ability, the *Trichoderma* strains grown on molasses-yeast medium at room temperature (28 ± 2 °C) is mixed with unsterilized soil, so as to get a conidial concentration of 10^6 per gram of soil, mix thoroughly in the soil and filled in plastic pots of 10 cm diameter. Matured, clean and polished paddy straw are cut into one cm segments and burying twenty pieces randomly in each pot. The pots are arranged in a completely randomized design covered with plastic cover to conserve moisture and incubate in the dark. All the 20 pieces from each treatment including control are to be removed from the pots after 7 days, washed in tap water to remove the adhering soil and debris. It is surface disinfected in a mixture of 1% sodium hypochlorite solution and 5% ethanol for 5 min. Segments are plated onto the TSM and incubated at room temperature (28 ± 2 °C) for 5 days. Per cent colonization of paddy straw pieces by the antagonist for each treatment has to be assessed. Three replications should be maintained. The CSA of *T. viride* mutants was assessed by the method of Ahmed and Baker (1987). Two CSA classes namely 1 and 2 (based on population mean of CSA, values less than population mean were assigned as 1 and that more than the population mean assigned as 2) are used to assess the antagonistic efficacy.

1.5.3 Colonization Behaviour of Antagonist on Pathogen (CB)

A modified form of the description of the antagonist based on their colonization behaviour can be used to group the *Trichoderma* strains. The modified description is as follows.

Table 1.1

Antagonism class	Description
01	Pathogen partially/completely overgrow the *Trichoderma* strain or colonized one third of medium surface and *Trichoderma* strain covered one third of medium surface.
02	Pathogen/*Trichoderma* colonized one half of the medium surface and neither dominated each other.
03	Initiation of overgrowth of antagonist on pathogen.
04	Over growth of *Trichoderma* on pathogen upto two third of the medium surface.
05	Complete overgrowth on pathogen and covered the entire medium surface.

The *Trichoderma* strains are said to be highly antagonistic, if the mean score for a given comparison is 3, but not antagonistic, if the mean comparison is between 1-2.

1.5.4 Percent Inhibition (PI)

Percent inhibition of the test pathogen by the antagonistic strains can be evaluated by dual culture technique. Two PI classes, namely 1 and 2 are (based on population mean of PI; values less than the population mean are assigned 1 and that more than the population mean assigned as 2) used to express the antagonistic efficacy. The sclerotial production of *R. solani* in antagonized plates can be counted visually using colony counter.

1.5.5 Propagule Lysis (PL)

The lysis of pathogen propagules can be assessed plating the pathogen grown under dual culture technique on PDA medium amended with 50 mg of carbendazim sensitive to *Trichoderma* strains but insensitive to pathogen. Five mm discs of medium per replication can be excised from the zone of interaction of antagonist and pathogen and plated on PDA. Based on the survival nature and lysis of the pathogen, a grading system of 0 to 5 can be used to express the propagules lysis.

1.5.5.1 Grading System for Propagules Lysis

Grade description of hyphal growth on agar surface

5 - No observable hyphae of pathogen

4 - Several threads or strands of hyphae

3 - Hyphae on 1-25% of the agar surface surrounding the disc

2 - Hyphae on 26-50% of the agar surface surrounding the disc

1 - Hyphae on > 50% of the agar surface surrounding the disc.

Two propagules lysis classes, namely 1 and 2 are assigned. If the mean score of a given comparison rounded to the nearest whole grade number in between 3 -5, then it is assigned as class 2 (highly antagonistic) and if the whole grade number falls between 1-2, it is assigned as class 1 (not antagonistic).

1.5.6 Speed of Overgrowth on Pathogen (SOOP)

The time taken by the antagonist/pathogen to overgrow after the contact of antagonist/both are assigned with four classes.

Table 1.2

Class	Description
01	Pathogens overgrow on *Trichoderma* strains
02	Neither antagonist / pathogen overgrow on each other
03	*Trichoderma* completely overgrow on pathogen after 48 hours
04	*Trichoderma* completely overgrow on pathogen within 24-48 hours.

The mean score of a given comparison falling between the classes 3-4 are antagonistic if the score falls between 1-2, then they are not antagonistic.

1.5.7 Inhibition Zone (IZ)

Besides the overgrowth on pathogen, a few strains formed a clear inhibition zone of varying thickness. The mean score of given comparison can be grouped into 4 classes (5-No IZ, 2-1 to 2.5 mm, 3-2.6 to 5mm, 4->5mm). For the mutant, that did not produce any inhibition zone, a weighted value of 1 can be given uniformly.

1.6 Factors Influencing Bio-efficacy

Abiotic factors influence the efficacy of biocontrol agents in soil. The factors include pH, moisture, temperature, soil type, components of soil atmosphere, inorganic and organic constituents of soil and the quantity and type of pesticides applied to soil.

1.6.1 pH

Hydrogen ion concentration plays a major role in the adaptability of biocontrol agents to their nutrient environment. Jeyarajan et al., (1994) found that the isolates of *T. viride* performed on a wide range of pH varying from 5.0 to 9.0, since the isolates of *T. viride* perform in wide range of soil pH, a composite, compatible formulation of *T. viride* could be developed and used in disease management programs.

3 Most of the pulses and oil seeds are grown in rainfed condition in different soil pH. Hence, isolates of *T. viride* suited to varying pH are to be identified and used. Culturing of *T. viride* strains 1, 2, 3, 6 in potato dextrose broth at different pH such as 5.0, 6.0, 7.0, 8.0 and 9.0 revealed that isolate 1 was able to grow at acidic and alkaline pH. But isolate 6 grew well at alkaline (pH-9). These variations indicated that strains specific to different soil pH could be selected and used for disease management.

Introduction of *T. viride* strains viz., MG6 and Tv. Wd as formulations into sterilized acidic and alkaline soils revealed that under sterile acidic conditions, multiplication of MG6 in soil was much greater in sorghum and paddy chaff based formulations. The population load was 7.0×10^6 on 30 days after incubation and declined to 5×10^4 cfu /g on 120 DAI in acidic sterile soils incorporated through sorghum-grain based formulation. It was followed by paddy chaff based formulation of MG6, and was not found to offer significantly from FYM-based formulation.

Colonization of acidic sterile soil by MG6 delivered through talc revealed that, the rate of proliferation was 8.2×10^2 cfu/g on 30 DAI and 1.8×10^2 cfu/g on 120 DAI, which was very much lower than that of the level required to control disease. Incorporation of Tv. Wd to acidic soil through inert carrier (talc) suppressed the population proliferation when compared to that of organic carrier (Nakkeeran et al., 2002). But survival of *T. viride* strains in alkaline soils was comparatively very poor than in acidic soils.

1.6.2 Soil Moisture

The soil moisture content is very important than soil reactions in producing changes in microbial flora in soil. An efficient antagonist should be able to survive in the prevailing soil moisture level. Isolates of *T. viride* and *T. harzianum* survived uniformly at 40% soil moisture level followed by 60% soil moisture. But 80 and 100% soil moisture were not favourable for the growth of *Trichoderma.* This indicates the suitability of the isolate to rainfed condition (Jeyarajan et al. 1994). Growth of *T. viride* was more at 40-60 per cent moisture than at higher soil moisture levels (Krishnamoorthy, 1987).

Soil application of *T. viride* as VBA–FB to soil having 40-60% soil moisture at 250kg/ha increased the seed germination and caused maximum reduction of root rot disease of mung bean. It recorded the maximum population of *T. viride* in rhizosphere soil and also increased the R:S ratio under field condition. Studies on population of *T. viride* strains in different soil moisture holding capacity showed a slow decline in the population level at >40 per cent soil moisture. But 100 per cent soil moisture was detrimental to all strains. Comparison of survival rate of various mutants indicates that MG6, MG3, MNT7 and UV 10 performed better than the parent strain at 40 to 80 per cent soil moisture. The same trend was followed after 60 days of incubation. But the population load declined from 10^5 to 10^4 cfu/g of soil.

1.6.3 Cellulase (CI)

The activity of cellulase by *T. viride* decides its potentiality to compete with other soil microflora. Hence the successful establishment of the antagonist depends on its competitive saprophytic ability for which the extracellular secretion of cellulose plays a vital role. Strains of *T. viride* like MG6, UV10 and MG3 had the highest activity of cellulose (6.1 units) followed by MNT7. It had a positive correlation with competitive saprophytic ability.

1.6.4 Hydrolytic Enzymes

1.6.4.1 Chitinase

The extracellular chitinase activity has been taken as a measure of antagonistic efficacy of *Trichoderma* spp. Chitinase activity of *Trichoderma* culture filtrates was enhanced when colloidal chitin was used as the sole carbon source. Chitinase pattern differed among the four selected strains of *T. viride*. The chitinase isoforms are induced differently by different carbon sources. The chitin affinity column fractions of *Trichoderma* culture filtrate inhibited the growth of *R. solani*. The extracellular chitinase of *T. viride* (TvMNT7) of molecular weight 42kDa was purified and its N-terminal sequence was determined. It had 70-100% homology with already characterized chitinase of similar molecular weight from other fungi. A chitin affinity column fraction that contained the enriched 42kDa chitinase showed a higher chitinase activity. It inhibited the growth of *R. solani* to the maximum extent under *in vitro* (Stephen Jebakumar et al., 2000). Specific activity of chitinase was maximum in colloidal chitin amended medium. It was 127.1 m moll of GlcNAc equivalent h^{-1} mg^{-1} protein for MG6. Among different cell walls as carbon sources, the mean chitinase activity was highest in the cell wall of sclerotial fungi than oomycetous fungi. The mean specific activity of MG6 in glucose-amended medium was 85.3, as against 94.2, 89.7 and 89.2 in the dialyzed culture filtrates of *R. solani, M. phaseolina* and *P. ultimum* cell wall-amended medium.

Chapter – 2

Mechanisms of Action of Biocontrol Agents

The antagonistic nature of *Trichoderma* as a biocontrol agent was first demonstrated by **Weindling** (1934), while the efficacy of *Trichoderma* under natural, field conditions was reported later by **Chet et al.** (1987). Further, more progress has been made towards the improvement of *Trichoderma* as a biological control agent in the last 15 years. Since then many potential strains of *Trichoderma* have been isolated and characterized from different natural habitat and used in biocontrol experiments against several plant pathogenic fungi.

Trichoderma uses a variety of mechanisms to provide protection against plant diseases. An **ideal antagonist** should possess characteristics such as, it may,

(i) directly kill the pathogen by mycoparasitism and/or antibiosis
(ii) adversely affect the growth and development of the pathogen either by antibiosis or by competing for nutrients, oxygen or space
(iii) alter the fitness of the pathogen
(iv) induce plant resistance
(v) enhance plant growth and its tolerance to stress
(vi) metabolize plant exudates supporting pathogen
(vii) inactivate enzymes produced by the pathogen.

The prevention of infection or suppression of disease by *Trichoderma* is based on **different modes of action** such as,

(a) Mycoparasitism
(b) Antibiosis
(c) Reduction of saprophytic ability
(d) Induced resistance and growth promotion
(e) Competition for nutrients and space
(f) Solubilization and sequestration of inorganic plant nutrients
(g) Inactivation of pathogen enzymes.

2.1 Parasitic Mechanism

The parasitic action includes mycoparasitism. It means parasitizing or overgrowing the hyphae of one organism by another organism there by inhibiting or suppressing the growth. Since its nascent stage, this has been one of the major mechanism involved in suppression of disease by *Trichoderma* (Weindling, 1932; Chet, 1987, Mukherjee et al. 1995, Elad, 1995, Howell, 2003 and Singh et al. 2003). This is a complex process involving tropic growth of biocontrol agent towards the target organism.

2.1.1 Mechanisms of Mycoparasitism

- Coiling of hyphae around the pathogen
- Penetration
- Formation of haustoria
- Dissolution of cell wall/cell membrane by the activity of enzymes

It also involves growth of *Trichoderma* hyphae along the hyphae of the host fungus, results in the formation of haustoria within the **large hyphae of *Rhizoctonia solani* by the smaller hyphae of *T. harzianum*.** There will be formation of loops, coil around, penetrate and colonize second stage larvae of the root-knot nematode, *Meloidogyne incognita*. Among species of *Trichoderma,* ***T. harzianum* exhibits excellent mycoparasitic activity against *R. solani*** where as ***T. virens* exhibit antibiosis against hyphae of same pathogen.** Hence, the nature of antagonism not only depends on antagonist but also on target pathogen. **Parasitism is more common against *R. solani*** while **antibiosis is the predominant mechanism of antagonism against fungi like *Sclerotium rolfsii* and *Fusarium* sp**. As far as mycoparasitism by *Trichoderma* is concerned, numerous genes and gene products have been proposed to be involved. Harman et al. (1989) tested the involvement of chitinase and β-1, 3 glucanase in *Trichoderma* mediated biological control by inactivating these two enzymes. The antagonist produces certain enzymes, proteases that inactivate the hydrolytic enzymes produced by the pathogen on leaves, as is evident in the case of biocontrol of *Botrytis cinerea* by *T. harzianum* on bean (Elad and Kapat, 1999). Biocontrol of root-knot nematode has been associated with protease production by *T. harzianum* (Sharon et. al. 2001). A transformed strain of *T. harzianum* (P2) containing multiple copies of protease gene reduced root galling better than wild type strains and it was the only strain able to penetrate eggs and egg masses. Regulation of individual genes presumably involved in mycoparasitism differs as reported by Zeilinger et. al. (1999). They showed that the **42- Kda endochitinase** is induced before *T. harzianum* comes into contact with *Botrytis cinerea,* while the **72- Kda N-acetyl hexosaminidase** is induced only after the two fungi are in direct contact. A single step in the mycoparasitic process of *T. harzianum* may involve **more than 20 separate genes** and gene products under complex regulatory control. Most of these products are synergistic with another. Given this entire arsenal of synergistic gene products that are part of only one mechanism by which *Trichoderma* species attack and gain nutrition from other fungi, it is not surprising that this genus has been reported to be against pathogenic, and provides control of very diverse groups of fungi (Chet, 1987; Harman, 2000). Studies on the molecular structure and characteristics of genes encoding enzymes of the chitinase complex in *Trichoderma* will contribute to a better understanding of the relationships of different enzymes involved in the biocontrol mechanisms.

2.2 Non-Parasitic Mechanism

2.2.1 Antibiosis

Trichoderma species produce different volatile and non-volatile compounds that inhibit the growth of fungal phytopathogens. The mechanisms of antibiosis constitute a much more complex system, leading to the phenomenon of biological control. Many antibiotics have been isolated and characterized from *Trichoderma*. These include **gliotoxin, gliovirin** and **glioviridin** that are extracted from ***T. virens*** (Sharma and Dohroo, 1991; Sivan and Chet, 1989, Howell, 2003). Besides, the population of a large variety of volatile secondary metabolites by *Trichoderma* e.g. **ethylene, hydrogen cyanide, alcohols, aldehydes and ketones** up to C4 chain length also play an important role in biocontrol. Howell and Stipanovic (1995) isolated an antibiotic, gliovirin, from *T. virens,* which was strongly inhibitory to *Pythium ultimum* and *Phytophthora* spp. but not to *R. solani*, *Theilaviopsis basicola*, *Phymatotrichum omnivorum* or *Verticillium dahliae*. The mutants deficient in gliovirin production lost the capacity to control damping-off of cotton. However, mutant with enhanced production of gliovirin was no more effective than wild type in controlling the disease (Howell, 2003). Lumsden et. al. (1992) demonstrated that disease-suppressive activity of *T. virens* against damping-off of zinnias caused by both *R. solani* and *P. ultimum* was correlated with production of gliotoxin by the biocontrol agent. The loss of gliotoxin production in *T. virens* mutants had an adverse effect on the efficacy of these strains as biocontrol agents of *Pythium* damping-off as reported by Wilhite et. al. (1994).

Another antibiotic compound, **Peptabiols,** which are non-ribosomally synthesized antimicrobial peptides, and exhibit antibacterial and antifungal properties represented by alamethicin (Reusser 1967; Landreau et. al. 2002) do play an important role in antagonism. To date, 309 peptabiols have been sequenced, among them more than 180 are synthesized by *Trichoderma.* Peptabiols form a class of antibiotics known for their high α-aminoisobuturic acid form and their synthesis is a mixture of isoforms ranging from 7 to 18 amino acids in length (Wiest et. al. 2003). Many peptabiols such as **trichozianins, trichokindins, trichorzins, trichorozins, harzianins, hypomuricins, trichoningins, trichokonins, trichogins** etc., exhibit a broad range of bioactivities related to cell membrane perturbations. They are thought to act on the membrane of the target fungus to inhibit membrane – associated enzymes involved in cell wall synthesis. Peptabiols have been shown to act synergistically with cell wall degrading enzymes to inhibit growth of fungal pathogens (Lorito et. al. 1996). This synergism is explained by an inhibition of the membrane bound β-1, 3- glucan synthase of the host fungus by the peptabiols, which inhibit the resynthesis of cell wall β-glucans, sustain the disruptive action of β-glucanases produced by *Trichoderma,* and all together enhance the fungicidal activity. Xiao-Yan et. al. (2006) studied antimicrobial metabolites produced by *Trichoderma koningii* SMF2 and showed them to be effective against a wide range of fungal phytopathogens. Although production of antibiotics by *Trichoderma* involved in biocontrol is a well-documented phenomenon, clear identification and understanding the role of antibiotics in disease control lags far behind than that in bacteria and it needs to be addressed (Whipps 2001; Singh et. al. 2002). Further research is needed to study the toxicity of antibiotic compounds and the mechanisms of their biocontrol activity against plant pathogens.

2.2.2 Competition and Rhizosphere Competence

(i) **Competition :** It is considered as classical mechanism of biological control. It involves competition between antagonist and plant pathogen for space and nutrients (Chet, 1987). It seems to be an important mechanisms of biocontrol, but it is difficult to assess its actual contribution in biological control. This mechanism is assumed to be involved if no evidence for mycoparasitism or antibiosis is found in a particular interaction (Alexander, 1982; Cook and Baker, 1983). The omnipresence of *Trichoderma* in agricultural and natural soils throughout the world proves that it must be an excellent competitor for space and nutritional source.

(ii) **Rhizosphere Competence :** It is the ability of antagonistic fungus to colonize rhizosphere. A biocontrol agent cannot compete for space and nutrients if it is unable to colonize the rhizosphere, a characteristic termed as **rhizosphere competence**. For a *Trichoderma* species to be rhizosphere-competent, it must colonize the rhizosphere beyond 2 cm depth from the seed (Chao et. al. 1986: Ahmad and Baker, 1987) or proliferate to a concentration that exceeds the initial population coated on the seed (Beagle – Ristaino and Papavizas, 1985). Rhizosphere competence of a particular isolate of *Trichoderma* makes it a successful biocontrol agent. *Trichoderma* species, either added to the soil or applied as seed treatment, grow readily along with the developing root system of the treated plant (Harman, 2000; Howell et. al. 2000; Zhang et. al. 1996). Although competition through rhizosphere competence may not be among the principal mechanisms that drive biological control, it is certainly a valuable adjunct to those that do. According to Harman (2001), **competition and rhizosphere competencies are the replacement of endogenous fungi on the root surface.** Possibly the ability of *Trichoderma* to colonize rhizosphere leads to increased levels of defense-related plant enzymes viz., peroxidases, chitinases and ß-1, 3-glucanases. In addition, root colonization by these beneficial fungi also induces significant changes in plant metabolic machinery. Biochemical analysis of sunflower, soybean, mustard and maize plants treated with *Trichoderma* formulations at different intervals of time after sowing shows substantial increase in the level of proteins, phenols, anti oxidants, total chlorophyll and vitamins in comparison to the un-inoculated control.

The information regarding biocontrol of *Botrytis cinerea* by *T. harzianum* strain T-39 was presented by Elad et. al. (1999). The conidia of *B. cinerea* require external nutrients for germination and infection. When conidia of T-39 were applied to leaves, germination of pathogens conidia was slowed, an effect attributed in part to competition (Elad, 2000). Thus, the competitive ability of *Trichoderma* and its biocontrol potential has shown to be affected by soil properties. Weindling (1934) noted that pH affected the relative value of *Trichoderma* as an effective antagonistic to *R. solani* by affecting the growth and development of each fungus. When *Trichoderma* is applied as seed treatment for the management of seedling and root diseases in chick pea, french bean and tomato, it is more effective than *Pseudomonas fluorescens* in acidic soil, whereas the latter is more effective in neutral and alkaline soil.

There is an upsurge in disease intensity when fungicides toxic to *Trichoderma* are applied for disease control. Backman et. al. (1975) showed that when fungicides toxic to *Trichoderma* were applied for peanut spot control, there was an increase in disease caused by *Sclerotium rolfsii*, a pathogen susceptible to biocontrol by *Trichoderma*. From this, it is evident that there was a wide array of factors that affect *Trichoderma's* competitive ability to be present and active as a biocontrol agent.

2.2.3 Siderophore Production

Siderophores are low molecular weight, virtually ferric-specific legends. They chelate Fe(III) and the microbial membrane receptor protein specifically recognize and take up the siderophore – Fe-complex (Neilands and Leong, 1986). This results in iron making unavailable to rhizosphere microorganisms, including plant pathogens, which produce less or different siderophores with low binding coefficients resulting in suppression of pathogen infection. Lorito et al. (1993) reported that *T. virens* produce three types of hydroxymate siderophores, namely,

(a) a monohydroxymate (Cis-and trans-fusarinines)

(b) a dipeptide of trans-fusarinine (dimerum acid)

(c) a trimer disdepsipeptide (coprogen).

2.2.4 Induction of Systemic Resistance (ISR)

It is now well-documented that treatment of plants with various agents viz., virulent or avirulent pathogens, non-pathogens, cell wall fragments, plant extracts and synthetic chemicals leads to the induction of resistance to subsequent pathogen attack, both locally and systemically (Walters et. al. 2005). The ability of *Trichoderma* spp. to induce local and systemic resistance has been shown with *T. harzianum* in agricultural crops such as bean, cotton, tobacco, lettuce, tomato and maize. Contrary to previously held opinions on biocontrol mechanisms, direct effects on plant pathogens are only one mechanism of biocontrol and are perhaps less important than induced resistance, according to Harman et al. (2004). With rhizosphere-competent strains that grow continuously with the plant, long-term systemic resistance is a possibility. The disease caused by natural infection of *Alternaria solani* was substantially reduced on foliage of sunflower and mustard by root application of *T. harzianum*. So, it indicate that induced resistance either local or systemic is an important component of plant disease control by *Trichoderma* spp. The molecular basis of resistance has been partially elucidated with the fact that *Trichoderma* strains produce a set of **ATP-binding cassette (ABC)** transporters (Lanzuise et. al. 2004). These include resistance to environmental toxicants that are produced by soil microflora or introduced by human activity and secretion of antibiotics and cell wall degrading enzymes that are necessary for establishment of a compatible interaction with a host of fungus. There are at least three classes of substances that elicit plant defense responses have been identified viz., peptides, proteins and low molecular weight compounds (Harman et. al. 2004; Viterbo et. al. 2005). The systemic response in plants occurs through jasmonic acid (JA)/ethylene signaling pathway in similar manner to the rhizobacteria-induced systemic resistance (Sarma et. al. 2002). These are proteins with enzymatic or other functions, homologues of proteins encoded by the avirulence (AVr) genes, oligosaccharides and other low molecular weight compounds that are released from fungal or plant cell walls by the activity of *Trichoderma* enzymes (Zeilinger et. al. 1999).

2.2.5 Growth Promoting Ability

It is evident from the experiments that formulations of *Trichoderma* significantly increased growth and dry weight of crop plants. Application of *T. harzianum* grown on wheat bran showed significant increase in shoot length, root length and dry weight of pepper, tomato and cucumber, whereas there is increase in height and flowering of chrysanthemum and periwinkle (Chang et. al. 1986). Increased fresh and dry weight of shoots, roots and nodules of broad bean were observed when the seeds were coated with *T. viride* (Yehia et. al. 1985). When rice seeds were treated with powdered formulations of *T. harzianum, T. virens* and *Aspergillus niger,* there was a significant increase in germination, seedling length and weight (Mishra and Sinha, 2000). Irrespective of antagonistic potential of all 43 isolates of *T. harzianum* tested against different soil-borne pathogens, there was significant enhancement in shoot growth of chilli, brinjal and pea. The growth promoting responses of *Trichoderma* spp. may be due to the following:

(i) Production of growth-stimulating factors such as hormones.

(ii) Suppression of deleterious root microflora including those not causing obvious disease.

(iii) Increased nutrient uptake through solubilization and sequestration of nutrients.

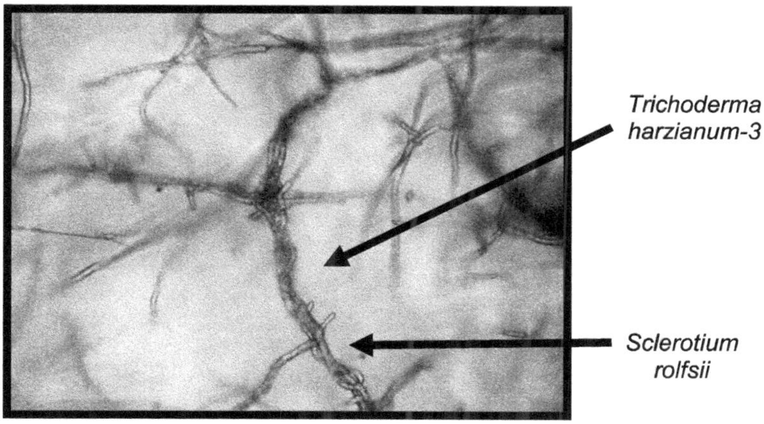

Fig. 2.1 Photomicrograph showing coiling of hyphae of *Trichoderma harzianum*-3 (Th-3) on Sclerotium *rolfsii* hyphae.

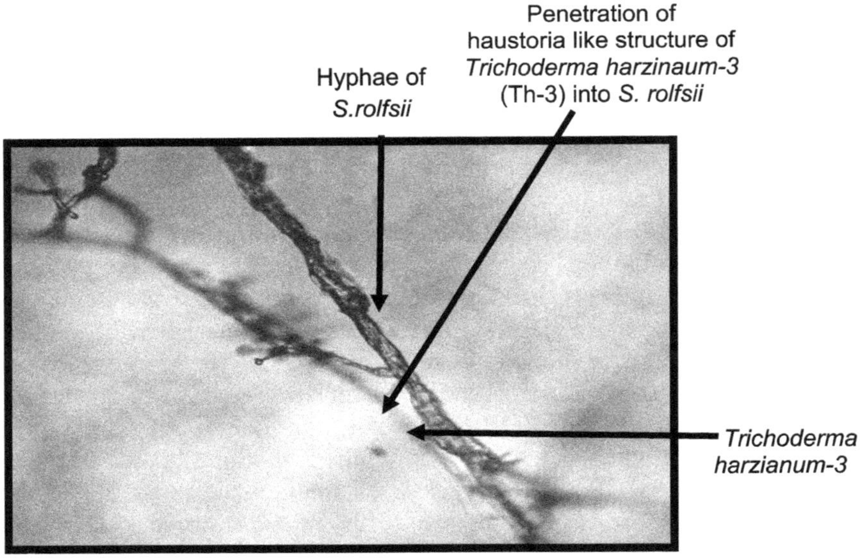

Fig. 2.2 Photomicrograph showing direct penetration of *Trichoderma harzianum*-3 (Th-3) into the hyphae of *Sclerotium rolfsii*.

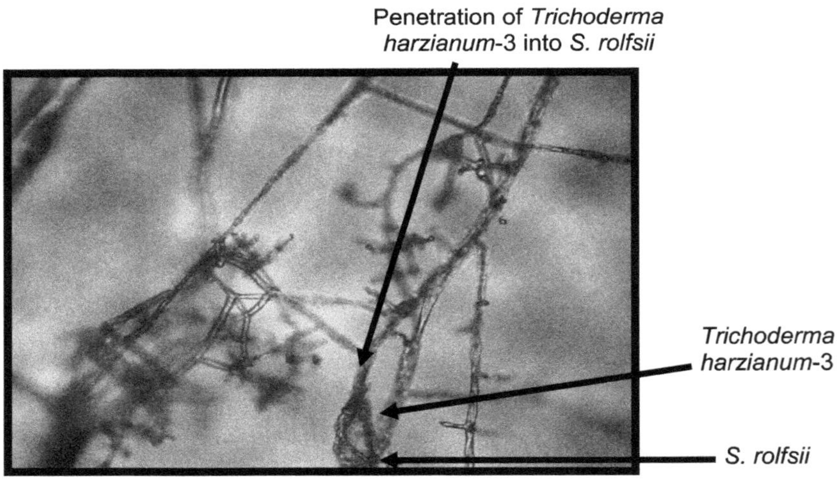

Fig. 2.3 Photomicrograph showing hyphae of *Trichoderma harzianum*-3 (Th-3) along the hyphae of *Sclerotium rolfsii* and penetration into *Sclerotium rolfsii.*

Chapter – 3

Selection of Potential Strain and Genetic Improvement of Bioagents

Isolation and screening of potential fungal biocontrol agents have been identified in numerous cropping systems or natural ecosystems (Baker and Cook, 1982). The first step in developing a fungal biocontrol agent is the discovery, through empirical or targeted screening, following isolation of potential strain of *Trichoderma* spp. for a particular or broad range of diseases. Screening of the fungal isolates for biocontrol activity is generally performed *in vitro*, along with tests to establish identity and relatedness of the newly found strains.

3.1 Important Steps Involved in Strain Selection

(i) Selection of effective strains in relation to plant pathogens.

(ii) Screening of isolates which have high biotechnological indices.

(iii) Analysis of pathogens properties specific for plants, useful insects, animals and humans.

(iv) Search for substrates which are economically viable and convenient for mass multiplication of fungus.

(v) Maintenance of high colony forming units (cfu) for longer period.

The relative ease of finding the right antagonist strain of *Trichoderma* as well as other biocontrol microorganisms under laboratory and field conditions is most often accomplished using various standard techniques available. This will be an added advantage but it does ensure that the agent would work consistently and effectively in field.

Bringing a biocontrol product to market is a multilayered process including discovery, efficacy trials, toxicological testing, mass production, formulation and registration (Singh et. al. 2001). The development of a highly successful strain T-22 of *T. harzianum* by Harman and his colleagues took more than a decade and it was only in later part of the last century that the sale of *Trichoderma*-based product started to pick up from scratch (Harman, 2000). Singh et. al. (2006) isolated several strains of *Trichoderma* of different species such as *T. harzianum*, T. *viride,*

T. atroviride and *T. koningii*. They have been found effective against large number of fungal phytopathogens affecting several economically important crops.

A potential strain of *T. harzianum* (ATCC-PTA 3701) useful as a nematode inhibitor, biofungicide and plant growth promoter (US Patent No. 6,475, 712) was developed by Karla et. al. (2002). The application of this strain as a biocontrol agent of soil-borne fungal pathogens had additional advantages of improvement of plant growth and economic yield of crop plants. It resulted in reduction of deleterious nematode population in the host tissue and rhizosphere thereby reducing the severity of root-knot disease and its use as soil amendment in reducing the application of hazardous chemical fungicides and nematicides which disturb the natural, beneficial soil microflora and pollute the soil and soil water. The application of this strain in nursery also reduced the input of chemical fungicides, which sometimes inhibit the rooting of cuttings. Similarly Singh and Singh (2004) screened the two strains viz., *T. harzianum* (IMI No. 359869) and *T. virens* (ITCC No. 1066.95) having the potential to control the collar rot disease of mentha species caused by *Sclerotium rolfsii* and also increase the oil yield significantly which was otherwise reduced drastically in diseased plants. It would be a boon to the industries associated with the development of biocontrol products, if such potential strains could be produced successfully in large scale and marketed for use by the farmers.

When once the identity of the fungus has been established, different isolates will be screened to assess diversity within the species. This will aid in selecting the most efficient and antagonistic strains for biocontrol ability *in vivo*. Thus, the ecological fitness is a fundamental requirement for bio-control agents (BCA's) because of the relatively narrow window of parameters particularly relative humidity, type of soil, osmotic potential and temperature. Adaptibility to these factors is a major requisite for *Trichoderma* strains to grow essentially in natural environment. However, isolates with the highest level of biocontrol *in vitro* may not perform as well *in vivo* since environmental conditions and competition with other microorganisms are much more restrictive. It is, therefore, essential to select fungal isolates under a range of conditions. Further, the efficacy of strains can be improved genetically.

3.2 Genetic Improvement of Strain

Biocontrol agents can be improved for desirable characteristics like better antagonistic ability, wider host range, tolerance to pesticides and survival ability in the environment, rhizosphere-competent and tolerant to adverse environmental conditions, vigorous growth and longer shelf-life. Different methods available for strain improvement are :

(a) Selection
(b) Mutation
(c) Protoplast Fusion
(d) Genetic engineering.

3.2.1 Selection

With respect to various characteristics of biocontrol agents such as antibiosis, competition for space and nutrients, growth promotion etc., a wide variability exists in natural population of different biocontrol agents. Majority of strains being used currently are selection from natural population. Existence of variability within the species can be further exploited to select efficient strains for different diseases, crops and environmental conditions. Natural biodiversity can also be exploited to target important diseases. Most of the commercial strains available today are selections from natural populations. Therefore, there is a need to collect, characterize properly and maintain as many isolates of biocontrol agents as possible.

3.2.2 Mutation

Mutation is the occurrence of sudden heritable change in genetic constitution of organism. The term ‘mutation’ is derived from a latin word ‘mutane’ means ‘to change’. Several chemicals and radiation are used to induce mutations. Chemical such as N-methyl–N-nitro-N-nitroso guanidine (NTG) has been most widely used for mutating biocontrol fungi. Ahmad and Baker (1988) generated mutants that were highly rhizosphere-competent and superior to wild types in respect of controlling *P. ultimum* in barley, radish, cucumber, pea and tomato and also in increasing fruit weight of cucumber by exposing the conidia of *Trichoderma* spp. to NTG. There are several other chemicals such as diethyl sulphate, ethyl methane sulphonate, sodium nitrate that are used to induce mutations. Liu (1988) used the above chemicals to induce benomyl resistant mutants of *T. koningii* which could tolerate up to 1000 mg/ml of benomyl and also without affecting antagonistic ability including lysis and antibiosis against *R. solani* and *S. rolfsii.* Production of more antifungal substances by mutants and was equally effective as the wild type strain in disease control potential was observed by Mukherjee et. al. (1997) by developing benomyl tolerant mutants of *T. viride* through chemical mutagenesis.

Several genetic variants of *T. harzianum, T. viride* have been developed using radiations such as UV rays, γ-rays etc. Benomyl tolerant strains of

T. viride were induced by exposing the conidia to gamma radiations (Troutman and Matejka, 1978). Application of ultraviolet (UV) radiations for the purpose of inducing mutations have been most extensive. Stable mutants of *T. virens* were developed by Mukherjee and Mukhopadhyay (1993) by exposing the cultures to 125k rad of gamma radiation. The resultant mutants differed from the wild type strain in phenotype, growth rate, sporulation and antagonistic potential. Increased production of biomass, abundant number of chlamydospores and conidia were obtained when *T. viride* strains induced with UV radiations at higher pH levels (Rajappan, 1997). A wide variation in enzyme activity especially β-1, 3-glucanase, β-1, 4-endoglucase and chitinase was observed in triazole – tolerant biotypes of *T. viride* developed by UV-irradiation (Kumar and Gupta, 2000).

3.2.3 Protoplast Fusion

Strain improvement of biocontrol agents can also be achieved by protoplast fusion. One of the commercially most successful strain of *Trichoderma* i.e. *T. harzianum* T-22 was developed by protoplast fusion of two strains i.e. T-95 which is a rhizosphere competent strain from columbian soil and T-12 from New York soil (Paulitz and Belanger, 2000).

Advantages

- Higher possibility of obtaining recombinants
- Testing of large number of recombinants in a short time
- This technology allows the induction of parasexual cycle at high frequency not only for the strains of same species but also for different species.

3.2.4 Genetic Engineering

It is an attractive option for the improvement of biocontrol strains as genetic modifications can introduce or improve desirable traits in fungal and bacterial antagonists. The enzymes in mycoparasitism such as chitinases, proteases, glucanases and the biosynthesis of these cell wall degrading enzymes is mainly controlled at transcriptional level (Dela cruz et. al. 1995) and responsible genes are present as single copy genes (Lora et. al. 1995; Limon et. al. 1995). Their gene copy number can be increased by transformation to overproduce these enzymes (Harman and Bjoorjman 1998). The resultant transformants over producing these enzymes are more efficient as biocontrol agents (Howell, 2003).

Chapter – 4

Molecular Characterization of Biocontrol Agents

Fungal genomics is becoming a major focus for research in the biological sciences that are vital for the development of new technologies for industry, agriculture and human health. Especially for genetic and biochemical research, fungi have become model systems that has elucidated the fundamental biology of eukaryotic organisms. The characters most often used in identification keys for filamentous fungi are based on arrangement, form, size and ornamentation of mature sexual and asexual structures. Another kind of latent characters used in identification keys are responses to abiotic factors such as macromorphology and colony growth rate or diameter at different temperatures, water activities, pH, redox potential etc. Detection of fungi on the basis of visual examination of morphology is highly selective and species-specific identification of fungi and spores is therefore difficult. The rapid development of high throughput DNA sequencing technology has provided a powerful tool for genetic research, from single gene cloning to whole genome sequencing. With the advent of technologies like DNA Microarray, it has been possible to study fungi on the molecular genetic level far more rapidly than ever could be achieved with traditional and biochemical genetic approaches.

Molecular characterization involves characterization of bioagents at protein and nucleic acid level and they were grouped into specific group or species level based on their identity. In the recent past, an increasing number of biocontrol agents are being developed for the control of insect pests and diseases in agriculture. It provides an immense source of data that can assist researchers in studying identity, genetic relatedness, diversity and selection of an ideal candidate for biological control. Further, it offers a means of constructing quality control tests that are essential throughout the developmental process of biocontrol agents. Selection of appropriate molecular techniques are based on the specific characteristics of the organism and on the desired type of information necessary to evaluate a particular step in the developmental process of a biopesticide. Genetic assessment of antagonists is not only desirable but essential to gain insight into the safe and reliable use of biological control.

4.1 Molecular Marker

It is a DNA sequence that is readily detected and whose inheritance can be easily monitored. The markers at DNA level include short DNA sequences, whole genes or even longer sequences of DNA. An ideal DNA probe would be known nucleic acid sequences that taxonomically define the target and present in high copy number.

Variation at genetic level can be detected by their expression in terms of different characters under varied environment. Transmission of characters from parents to progenies, through genetic material and expression of those differentiating characters in the progenies are called **genetic markers**. Three types of markers can detect the genetic differences at chromosomal level.

4.1.1 Morphological Markers

Different plant characters like plant height, leaf shape and pathogen characteristics like culture growth, pigmentation, pathogenicity and other visual identifiable characters are known as morphological markers.

4.1.2 Biochemical Markers

These include, analysis of proteins and enzymes like isozymes or allozymes extracted from the tissues. Proteins and isozymes are used as markers after distinguishing the polymorphisms in the electrophoretically separated and specially stained protein bands.

4.1.3 DNA Markers

Analysis of DNA in the form of individual bands or restriction fragments is the direct method of estimating large number of differences at genetic level.

Morphological and biochemical markers are the ultimate result of gene expression, which are influenced by the environment, as well as developmental stages of the organism under study. Besides, these markers are few in numbers and do not represent the whole genome of an organism, and hence cannot be used for the construction of a saturated genetic map. However, DNA markers are enormous in numbers and stably inherited in Mendelian fashion. These are also not affected by environment or developmental stages of the plants. Variation at DNA level is detected by DNA polymorphisms survey by various molecular techniques. Polymorphisms at the DNA level are the result of simple point mutation, insertion or deletion of DNA segments etc.

4.2 Types of DNA Markers

Natural variation at DNA level can be detected by a range of molecular, biological techniques developed in the recent past. The choice of their use solely depends on nature of the study and crop species in question. Different DNA markers have been classified into three categories which are explained below.

4.2.1 Hybridization-based Markers

The most important hybridization-based technique is RFLP. In this technique, DNA is first digested with restriction enzymes, separated on an agarose gel and further transferred to membrane filter. DNA fixed on the filters is then hybridized with locus-specific DNA probes labeled with radioactive or non-radioactive substance and DNA fingerprints are obtained by autoradiography. Alternatively, hybridization can be performed with micro satellite specific DNA probes. Such types of DNA markers are then termed as variable number tandem repeats (VNTR) and oligonucleotide fingerprinting.

4.2.2 Sequence Targeted and Single Locus PCR-based Markers

Through DNA sequence, information can be rapidly detected by the use of DNA markers. Some of the PCR-based DNA markers do not provide allelic information. Such problems can be overcome by the use of PCR primers flanking to a specific locus. Most ecological and evolutionary studies are based on DNA markers derived from the sequences of DNA. However, it is a labour-intensive and expensive technique and sometimes difficult to use in the detection of polymorphisms at species or race level. Thus PCR-based technique like cleaved amplified polymorphic sequence (CAPS) can prove better. The technique, which provides useful easy-to-screen markers, is also known as PCR-RFLP, since, in this technique PCR products are first digested with restriction enzymes before their electrophoresis on agarose gels. For the specific and sensitive detection of locus, other PCR techniques like allele-specific oligonucleotide ligation assay (OLA) and other cloning and sequencing of simple sequence repeats (SSR), PCR primers flanking to the region can be produced. These are known as sequence-tagged microsatellite site (STMS) markers.

4.2.3 PCR-based Markers

Amplification-based DNA markers have been proved versatile and easy because there is no need of difficult and lengthy steps of Southern blotting. All the genetic markers based on PCR are generated by the modifications of original method of PCR developed by Prof. Karry Mullis and his group of the United States of America in 1985. The technique is very sensitive and allows specific amplification of DNA fragments from the genomic DNA of all the organisms. PCR is a simple chemical synthesis of DNA molecule with the help of DNA polymerase and primers by using genomic DNA as template. The technique can be automated for large-scale applications and devoid of radioactivity used for labeling DNA probes. Some of the PCR-based markers are arbitrary primed PCR (AP-PCR), amplified fragment length polymorphism (AFLP), random amplified polymorphic DNA (RAPD), multiple arbitrary amplicon profiling (MAAP), DNA amplification fingerprinting (DAF), arbitrary amplified DNA (AAD), simple sequence repeats (SSR), and RNA arbitrary primed PCR (RAP-PCR).

4.3 Non-PCR-based Techniques

4.3.1 Restriction Fragment Length Polymorphism (RFLP)

It is simple southern blot analyses mainly used to detect specific types of sequences given in a system based on specific probe. It was the first technology that enabled the detection of polymorphism at DNA sequence level.

Principle : The occurrence of variation in the length of DNA fragments that are produced after cleavage with restriction endonuclease. The differences in the DNA lengths are due to the presence or absence of recognition sites for that particular restriction enzyme.

Method : RFLP analysis comprises the following basic steps.

(a) DNA isolation.

(b) Cutting DNA into smaller fragments using restriction enzymes like ECoRI.

(c) Separation of DNA fragments by gel electrophoresis.

(d) Transferring DNA fragments to a nylon or nitrocellulose membrane filter.

(e) Hybridization with labeled probe to DNA fragments.

(f) Visualization of specific DNA fragments on X-ray film.

4.3.2 Dot-Blot Hybridization (DBH)

It is the hybridization-based assay most commonly used for pathogen detection.

Method : It comprises the following basic steps.

(a) Crude extract containing DNA from infected samples is taken and spotted onto the nitrocellulose sheets.

(b) Denaturation of DNA, neutralization and baking at 80 °C for 2 hrs.

(c) Pre-hybridization and hybridization with a labeled probe is allowed and detected on X-ray film which gives the positive indication of presence of pathogen.

Probes can either be developed by random cloning of genomic DNA of pathogen or its cloned fragment involved in toxic production pathogenicity.

Variability among variants of mungbean mosaic gemini virus (MYMV) by nucleic acid spot hybridization was reported by Kajal Kumar Biswas and Anupam Varma (2000).

4.4 Gel Electrophoresis

Sailaja Rani and Reddi Kumar (2007) studied variability among six isolates of *Pythium aphanidermatum* collected from three districts of Rayalaseema region by subjecting to protein and isozyme analysis. Significant variation was observed among the six isolates with respect to protein and esterase patterns. SDS-PAGE profiles of proteins indicated that an isolate, CTR2 was unique in having one additional band with a molecular weight of 8 kda. This protein band was absent in the remaining five isolates. Native PAGE profiles of esterase enzyme revealed the absence of a band at Rm 0.40 in CTR2. (Fig. 4.1)

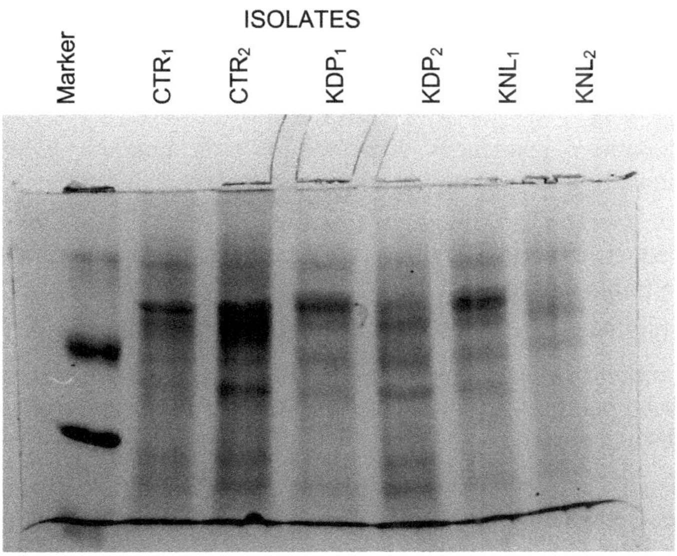

Fig. 4.1 SDS-PAGE analysis of total fungal proteins in six isolates of *P. aphanidermatum*.

4.5 Polymerase Chain Reaction (PCR)

This technique was first developed by Kary Mullis et. al. (1988). It is an *in vitro* method for the amplification of DNA fragments referred to as polymerase chain reaction. The amplification of DNA is achieved by a repetitive series of cycles involving three steps.

Important Steps

(a) Denaturation of a template DNA duplex by heating at 94 °C.

(b) Annealing of oligonucleotide primers to the target sequences of repeated DNA strands at 35-65 °C.

(c) DNA synthesis from the 3'-OH end of each primer by DNA polymerase.

The key to success of PCR as a diagnostic tool lies in the design of pathogen specific primers.

- Primers exhibiting specificity at the generic, species or race level can be designed from the sequences that are unique to the organism.
- One source of such sequences for bacterial pathogens is in their plasmids.

4.5.1 Random Amplified Polymorphic DNA (RAPD)

RAPD analysis is a PCR-based molecular marker technique which utilizes single short oligonucleotide primers of arbitrary sequence for the amplification of segments distributed randomly throughout the genome.

Important Steps

(a) Repetitive sequences dispersed in total genomic DNA provide annealing sites of universal primers.

(b) As a result of binding of primer at random to complementary sequences, it will allow the amplification of several bands of variable length DNA.

(c) Fragments can be resolved using gel electrophoresis.

Gilly zimand et. al. (1994) used RAPD procedure for identification of *Trichoderma* strains. 47 *Trichoderma* strains were analyzed by RAPD procedures using primers 1, 3 and 4.

Strains of *Pseudomonas putida* were identified by PCR amplification and direct sequencing of gyr B genes with universal primers. (Satoshi Yamamoto and Shigeaki Haryama 1995.)

Abbasi et. al. (1999) used RAPD – PCR technique to develop markers for reliable differentiation of *T. hamatum* 382 in compost-amended potting mix. A total of 180 random primers were secured against purified genomic

DNA of 10 randomly selected isolates of *T. hamatum* and three primers OPH-19, OPE-16, OPH-20 were selected and by using these three primers 53 isolates of *T. hamatum* were differentiated.

RAPD technique was used to develop a genus-specific probe for *Streptomyces* and a strain – specific probe for the biological control agent, *Streptomyces lycidus* WYEC 18 (Mark. A. Roberts and Don L. Crowford 2000). Lih-ren Kong et. al. (2001) characterized a Taiwanese strain of *Streptomyces saracetious* N45, exhibiting high chitinase activity using RAPD.

Two isolates each of six species of *Trichoderma* (*T. virens, T. pseudokoningii, T. hamatum, T. harzianum, T. viride and T. koningii*) were characterized by Latha et. al (2002). Two species of *Trichoderma* (*T. reesii* and *T. harzianum*) were differentiated at molecular level using RAPD technique by Nagee et. al. (2003). Reddy et. al. (2004) characterized *Pseudomonas fluorescens* strains producing antifungal compounds by PCR and RAPD analysis.

Venkateswarlu and Reddi Kumar (2007) studied genetic variability among the isolates of *Trichoderma* spp. using molecular techniques like RAPD and ITS-PCR. The RAPD banding pattern reflected the genetic diversity among the isolates with the formation of 7 clusters (Fig. 4.2). ITS region of rDNA amplification with genus specific ITS 1 and ITS 4 universal primers produced product size varies from 504 to 566 bp in all the isolates indicated polymorphism at ITS region of rDNA among the *Trichoderma* isolates (unpublished).

4.5.2 Amplified Fragment Length Polymorphism (AFLP)

It is a novel DNA finger-printing technique. It involves display of a set of DNA fragments from a specific DNA sample. This is highly sensitive method for detecting polymorphism throughout the genome.

In principle, it involves the following steps.

(a) DNA is cut with two restriction enzymes.

(b) Double stranded (ds), oligonucleotide adapters are ligated to the ends of DNA fragments.

(c) Selective amplification of restriction fragments with two primers according to the sequences of adapters plus 1-3 additional nucleotide fragments contain restriction site sequence plus the additional nucleotides will be amplified.

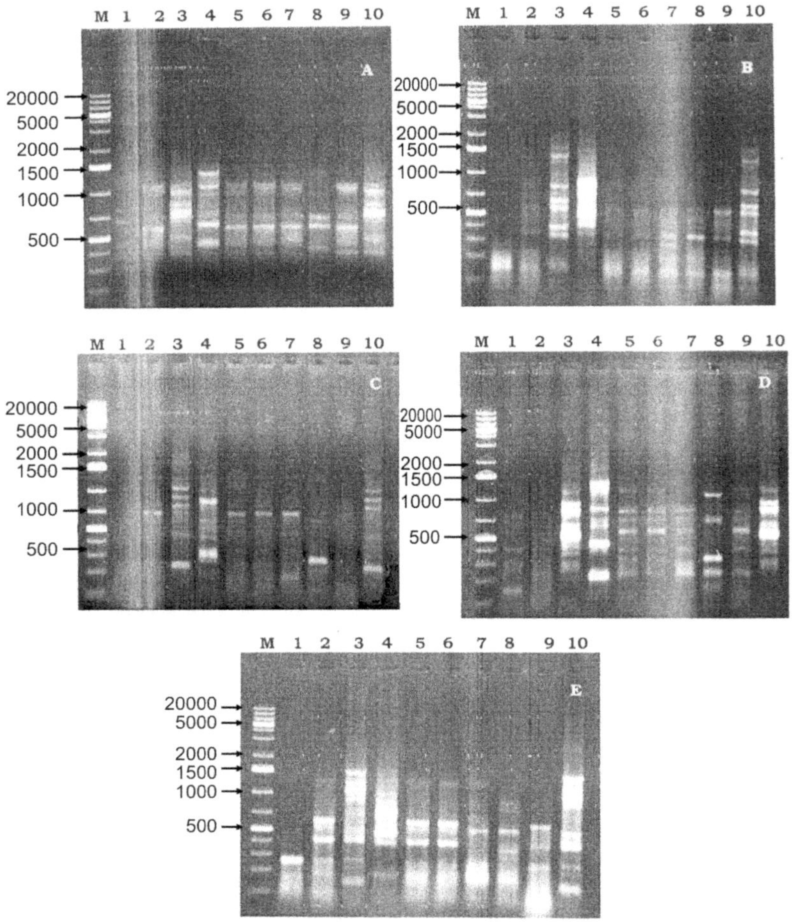

Fig. 4.2 Random amplified polymorphic DNA (RAPD) patterns of *Trichoderma* isolates with various random primers; A : OPA-08; B : OPA-09; C: OPA-10; D: OPA-12& E: OPA-13; M : Molecular weight marker of 1 kb DNA ladder. Lanes 1 to 10 are T1 to T10 *Trichoderma* isolates, respectively.

(d) Gel analysis of the amplified fragments. The amplification products are separated on highly resolving sequencing gels and visualized using autoradiography.

The number of AFLP polymorphic bands produced in a single reaction is determined by the specificity of restriction enzyme primers used to digest the genomic DNA, the number and choice of selective nucleotides at the 3' end of the restriction enzyme primers. Genetic variability among 36 isolates of *Fusarium udum* collected from pigeon pea growing states of India was studied by Siva Ramakrishnan et. al. (2002) using RAPD and AFLP technique.

4.5.3 Sequence Characterized Amplified Regions (SCAR)

A SCAR is genomic DNA fragment at a single genetically defined locus that is identified by PCR amplification using a pair of specific nucleotide primers. Paran and Michelmore (1993) converted RAPD markers into SCAR. In order to develop a rapid, simple and specific diagnostic PCR assay for *T. hamatum* 382, the three RAPD markers (OPH-19, OPE-16, OPH-20) were converted into SCAR markers (Abbasi et. al. 1999). Rubio et al. (2005) generated sequence characterized amplified region (SCAR) markers to identify the biocontrol strain *T. harzianum* 2413.

4.5.4 Cleaved Amplified Polymorphic Sequences (CAPS)

In this technique, PCR-amplified DNA fragments are digested with restriction enzymes to reveal restriction site polymorphisms. This is also some times known as PCR-RFLP. The digested restriction fragments are subjected to agarose gel electrophoresis followed by staining with ethidium bromide for visualization of bands.

Lorenz et. al. (1995) analyzed ribosomal sequences to design PCR primers for detection and identification of the phytoplasmas that cause apple proliferation and pear decline.

4.5.5 Internal Transcribed Spacer- PCR (ITS-PCR)

In ribosomal DNA, three sub units namely 28s, 18s and 5.8s are joined with spacer 1 and spacer 2. By ITS – PCR technique, ITS region is amplified by universal forward and reverse primers. ITS-based identification can be based on the number and length of the PCR-amplified products. Hermosa et. al. (2000) used ITS –PCR and RFLP and identified 16 biocontrol strains of *Trichoderma* into *T. harzianum – T. inhamatum* complex (9), *T. longibrachiatum* (1), *T. asperellum* (3) and *T. atroviride – T. koningii* complex (3) based on ITS region.

4.6 Advantages of Molecular Techniques

- Molecular techniques do not rely on phenotypic examination.
- Expression of gene is not required and thus identification time can be reduced.
- Nucleic acid sequences unique to particular fungi can be generally found.

- Some methods have the potential for the detection of specific genomic DNA sequences directly from initial plant samples, thereby eliminating the requirement to isolate and culture the fungus.
- Specific molecular markers, probes and primers have been developed from a variety of DNA sequences.
- In the case of antagonistic fungi, molecular markers can be derived from both variable and conserved regions of nuclear and mitochondrial genome.
- Different markers can be used to define populations at all levels from an individual isolate upwards.

Chapter - 5

Morphological Characteristics of *Trichoderma* spp.

Trichoderma a genus of asexually reproducing fungus, is present in nearly all temperate and tropical soils. The strains of *Trichoderma* spp. are strong opportunistic invaders, fast growing, prolific producers of spores and powerful antibiotic producers. These properties make these fungi ecologically very successful and are the reasons for their ubiquitousness. They show a high level of genetic diversity, and can be used to produce a wide range of products of commercial and ecological interest. Much of the known biology and many of the uses of these fungi have been documented recently (Buhariwalla et al 2005; Druzhinina 2006). Some strains of *Trichoderma* are highly rhizosphere competent i.e. able to colonize and grow on roots.

Trichoderma spp. are the most extensively used fungal biocontrol agents for the management of plant pathogens affecting seed, root and aerial plant parts. The parasitic nature of *Trichoderma* spp. was first reported by **Weindling (1932)** on several plant pathogens, including *Rhizoctonia solani, Phytophthora parasitica, Pythium* spp. and *Sclerotium rolfsii.*

5.1 Exploitation of *Trichoderma* on Commercial Scale

They are highly efficient producers of many extracellular enzymes like cellulases, chitinases, glucanases, proteases etc.

(a) It form a source of cellulases which can be used in foods, textiles and poultry feed.

(b) The enzyme chitinase would form an important source of generating disease-resistant transgenics.

(c) It plays an important role in plant disease control through antifungal and anti-nematode activity and on plant defense induction.

(d) It helps in improvement of plant growth.

(e) It augments the process of straw or compost decomposition and also in suppression of weeds.

5.2 Taxonomic Status

Trichoderma produces only asexual spores or conidia and hence can be classified as imperfect fungi. *Hypocrea*, an ascomycete found to be perfect state of the genus. Rifai (1969) has outlined the speciation concept within the genus *Trichoderma* and described nine species aggregates which were based on **morphological characters**. However, with the use of molecular approaches, particularly sequencing of internal transcribed spacer-1 (ITS 1 and ITS 2), the taxa recently have gone from nine to 35 species.

The genus is characterized by rapid growth rate in culture and production of numerous single celled spores (conidia), which have various shades of green. Conidiophores are produced in tufts with divergent but often irregularly bent flask-shaped phialides. Conidiophores may end in sterile appendages with the phialides only borne on lateral branches in some species. The reverse side of colonies is often uncoloured, amber or yellow green. Chlamydospores are often produced by many species in submerged mycelium. Rifai (1969) differentiated *Trichoderma* from *Gliocladium* primarily the **angle at which branches and phialides are borne**. In *Gliocladium* branching **is penicillate or converged with phialides closely appressed.** While the **phialides are straight** and **moderately divergent,** those cultures are placed in **Verticillium.**

Trichoderma cultures are best grown on 2% malt extract agar in day light or under near UV light. Very young cultures, usually five days old, can be employed for examining phialides structure, whereas texture of conidia can be best studied in two weeks old cultures.

5.3 Morphological Characteristics of Important *Trichoderma* spp.

5.3.1 T. harzianum Rifai

Conidia smooth walled, subglobose to obovoid, 2.8 to 3.2 × 2.5 to 2.8 um, phialides regularly disposed in numbers of 3 or more, not crowded conidiophores and their side branches are long and slender without sterile hyphal elongation. Colonies yellowish, bright to dark green, floccose or with compact conidiophore tufts. Colony reverse uncoloured. Colonies cover entire Petri plates (90mm) within 4 days at 25 ± 2 °C on PDA. (Fig. 5.2)

5.3.2 *T.viride* pers. Ex Gray

Conidia with rough surface single celled, green or brownish, subglobose to ellipsoid. Phialides mostly in virticils of 2 or 3 are not crowded.

Conidiophores are narrow, flexuous, primary branching at regular intervals, paired or in whorls of three, short and are not extensively branched. Chlamydospores are present in most isolates. Colonies yellowish, bright dull to dark green, are grown rapidly and cover the Petri plates after 4 days at 20^{0}C on oat meal agar medium.

5.3.3 *T. virens* (= *Gliocladium virens* Miller, Giddens & Foster)

Conidia dark green are broadly ellipsoidal to obovoid conidia from adjacent phialides and often coalesce into large globoid masses, phialides broadly ampulliform. Fast growing hyaline colonies bear repeatedly branched conidiophores in tufts with divergent, often irregularly bent, flask shaped phialides. Conidiophores may end in sterile appendages with the phialides only borne on lateral branches. In some species colonies spread rapidly, aerial mycelium floccose, white to greyish. Colony reverse is colourless. (Fig. 5.1)

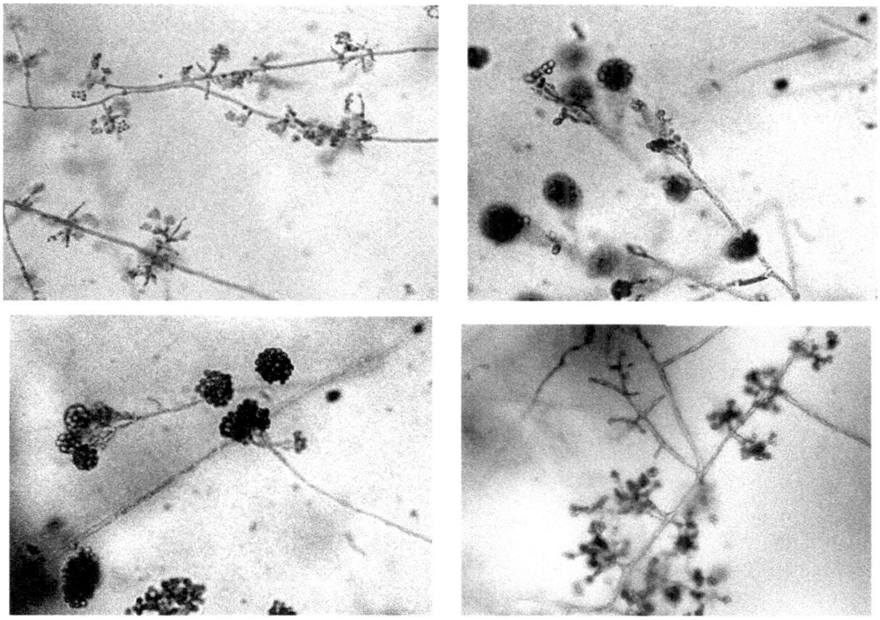

Fig. 5.1 Microscopic features of *Trichoderma virens (Gliocladium virens).*

5.3.4 *T. hamatum* (Bon.) Bain

Conidia are smooth walled or finely punctuated, green in colour, oblong to ellipsoid. Conidiophores are long and thick, often with sterile hyphal elongations, side branches mostly short and thick bearing crowded short and plump phialides. Colonies are white or whitish green to green, generally with compact conidiophore tufts. Chlamydospores are abundant, colonies that grow moderately rapidly, limited, and aerial mycelium is floccose.

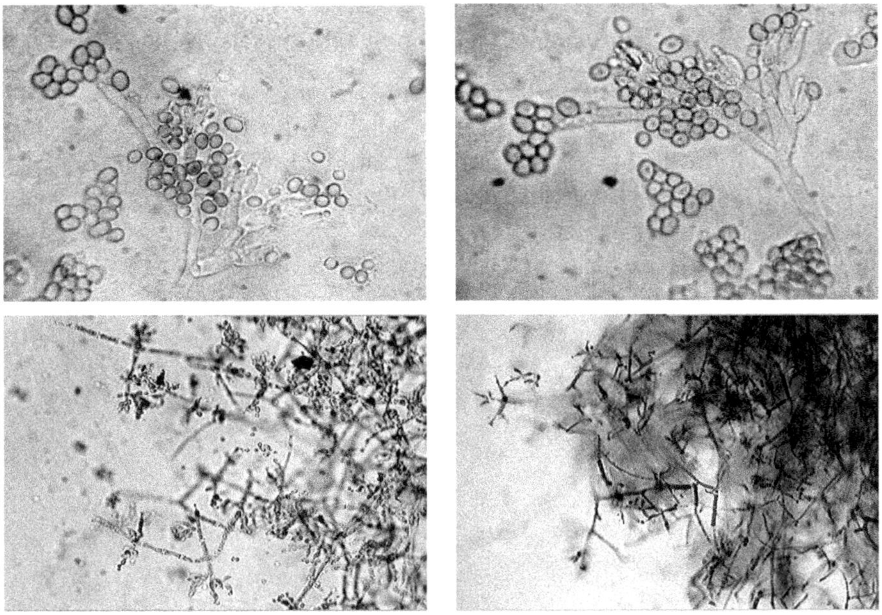

Fig. 5.2 Microscopic features of *Trichoderma harzianum.*

Chapter – 6

Mass Production and Formulation of *Trichoderma* spp.

Trichoderma can be mass produced using solid substrates and different liquid media. Most of the agricultural wastes or byproducts can be used as substrates for the intended purpose. The inoculum of *Trichoderma* should survive during each processing step such as harvesting, drying, formulation, storage and delivery. Production of inoculum is rather time bound and should be cost-effective. Based on nature of substrates used for inoculum production, two major methods are available, namely,

- Solid state fermentation
- Liquid state fermentation

6.1. Solid State Fermentation

This is a very common method used for mass production of *Trichoderma* spp. Various cheaply available cereal grains like sorghum, ragi, bajra, millets etc., can be used as substrates. For this, the grains are soaked in water overnight and water is decanted in the next morning. The soaked grains are taken in polypropylene bags and sterilized for 30 minutes in an autoclave or pressure cooker. The sterilized grains are inoculated with *Trichoderma* spore suspension and incubate for 10-15 days. *Trichoderma* produces dark green spore coating on grains. These grains are powdered finely and can be used for seed treatment or for enriching farm yard manure (FYM) for soil application.

6.2 Substrates used for Solid State Fermentation

6.2.1 Sorghum Grain-based Formulation

It is inexpensive and cheaply available for multiplication of inoculum in large scale. The speed of colonization of the substrate varies from 7 to 13 days. Strains of *T. viride* MG6, MG3, UV10, MNTT, UV2 and Tv. Wd were formulated using sorghum grain as substrate (Nakkeeran, 2001).

6.2.2 Vermiculite – Wheat bran

Vermiculite	:	100g
Wheat bran	:	3.3 g
Liquid culture	:	14 ml
0.05 N HCl	:	17.5 ml

The components are mixed proportionately and a product with wheat bran as base was formulated by Vidya (1995) and applied to soil at the rate of 25 kg/ha to mungbean. There was 41% reduction in root rot incidence caused by *M. phaseolina* and 91% increase in yield. Now this product is commercially prepared as 'Antagon combi' and widely used by planters in Karnataka at the rate of 10 kg/ha to manage pepper root rot. Vermiculite bran acid fermentor biomass (VBA-FB) of *T. viride* recorded the highest mean population of 20.5 $\times$ 10^7 cfu/g upto 75 days of storage at room temperature.

6.2.3 Press Mud

Press mud is available as a byproduct of the sugar factory and can be used as a substrate for mass multiplication of *Trichoderma* and sold to farmers as value-added organic manure. In other sugar factories, press mud is disposed of by permitting farmers to lift it to their fields after delivering sugarcane to the factory for crushing. This unique process has converted waste to wealth.

6.2.4 Mushroom Spent Waste-based Formulation

The spent waste of paddy straw (10kg) after growing the oyster mushroom (*Pleurotus*) is mixed with decomposed coir pith (2.5kg) and the moisture adjusted to 50% with 10% jaggery solution. Then the mixture has to be taken in polypropylene bags (400 g /bag) and sterilize at 15 1b pressure for 1 hr. and inoculated with 9 mm disc of *Trichoderma*. The bags are to be kept for incubation at room temperature (30-32 °C) for 8-10 days. By this method, spore load reached 40 $\times$ 10^9 cfu /g after 90 days of incubation (Nakkeeran, 2001).

6.2.5 Paddy Chaff-based Formulation

By using this substrate, a very high spore load of 22 $\times$ 10^4 cfu/g was recorded. For this, paddy chaff was boiled in 10 per cent molasses / jaggery solution (1:2) for 30 min. After decanting the molasses solution, shade dried over a gunny bag for 15 minutes. To that 10 g of $CaCO_3$ was added (20g/kg) to get a uniform coating. The chaff was packed in autoclavable polypropylene bags (200 g/bag) and sterilized at 15 1b pressure for 45 min.

To the sterilized substrate 5-day old, 9 mm discs of *T. viride* are inoculated and incubated at room temperature for 25 days (Nakkeeran, 2001). It can be used as primary inoculum for *in situ* multiplication in FYM or decomposed coir pith at the rate of 1 kg for 100 kg.

6.2.6 Decomposed Coir Pith

Coir pith is available aplenty as a waste of coir industry and can be converted into manure by *Pleurotus*. *Trichoderma* was also as effective as *Pleurotus* in decomposing coir pith (Vidya, 1995). Besides, such *Trichoderma*-decomposed coir pith also served as *Trichoderma* product when applied to soil.

6.2.7 Coffee Husk

Coffee husk is a waste from coffee curing industry. Based on this, a *Trichoderma* formulation was developed, which was found to be very effective in managing *Phytophthora* foot rot of citrus (Sawant and Sawant, 1996) and is widely used in Coorg, Karnataka state.

6.2.8 Other Substrates

Farm yard manure, gobar gas slurry, rice bran and groundnut shell are also found to support the growth of *T .harzianum* and *T. viride* (Kousalya and Jeyarajan, 1988).

6.2.9 Limitations of Solid State Fermentation

- As the preparations are bulky, they may be subjected to a greater risk of contamination.
- Require extensive space for processing, incubation and storage.
- As these difficulties make the system inappropriate for commercial development.

6.3 Liquid State Fermentation

At present, this system appears feasible to produce large quantities of inoculum of various beneficial microbes. By this process, desirable quantity of biomass can be produced within a few days. Studies at PDBC, Bangalore revealed that maximum amount of biomass and viable propagules of *T. harzianum* can be obtained within 96 h of fermentation in a fermentor with aeration, agitation and temperature control (Prasad et. al. 1997). Maximum biomass of *Trichoderma* spp. can be realized in a short time by using appropriate medium in a fermentor with aeration, agitation, temperature, pH and antifoam controls than in shake-flask cultures. It will be more suitable for industrial production of *Trichoderma* spp. In industrialized countries deep tank fermentation in inexpensive media such as molasses and brewers' yeast for large scale production of *Trichoderma* is being followed.

6.3.1 Liquid Media Used

(i) Jaggery Soya Flour Medium

Jaggery	-	50 g
Soya flower	-	10 g
Water	-	1 lit.

(ii) Molasses Yeast Medium

Molasses	-	30 g
Yeast granules	-	5 g
Water	-	1 lit

(iii) Potato Dextrose / Jaggery Medium

Potato	-	200 g
Dextrose / Jaggery	-	20 g
Water	-	1 lit

For obtaining optimum spore load or inoculum, the antagonist can be grown in any one of the liquid media. Based on function, three types of culture conditions are available such as (a) Stationary (b) Shaker (c) Fermentor. In stationary culture condition, it will take about 10 days for full spore production. While under shaker culture conditions, the antagonist requires 7 days, in fermentor, peak production can be achieved after three days.

Before a biological control agent (BCA) can be successfully introduced into the market, a series of studies must be carried out. Concomitantly, production and formulation of an inoculum with an acceptable shelf-life have to be achieved, which is frequently becoming a major bottleneck. With the availability of large quantities of an inoculum, field and glass house tests on relevant crops are to be done to determine the reproducibility of control. Until recently, the commercialization of *Trichoderma* spp. as a biocontrol agent has been hampered by the lack of cost-effective way of producing sufficient amounts of fungal material to be used in field and glass house trials.

Chapter – 7

Formulation and Shelf-Life of *Trichorderma* spp.

Formulation is the blending of active ingredients with inert carriers, plant materials or inert materials. The best technology is of no value unless it can be placed easily in the customer's hand for use. Hence, developing a safe, easy to use, cost-effective formulation that will keep the *Trichoderma* alive is one of most important steps in developing a final product. In the case of *Trichoderma* formulation, the active ingredient consists of *Trichoderma* propagules and hyphae with inert carriers like talc, peat and peat plus additives, coal and coal plus additives, inorganic soils etc. Plant materials comprised of compost made from bagasse, saw dust, soybean, coir dust, rice husk, plant compost, FYM powder etc., or inert materials like vermiculite, vermiculite plus additives, perlite, ground rock phosphate, calcium phosphate etc., can be used. Carboxy methyl cellulose (CMC) @ 1% is used with talc powder as an adhesive. Modern formulations of bioagents are liquid-based. They are made usually in mineral oil along with emulsifiers.

For the successful development of formulation of biocontrol agents, it is important not only to provide a substrate that will promote the synthesis of the desired enzymes which help in its biocontrol mechanism but also to provide sufficient substrate so as not to limit the synthesis of enzymes at the time they are required. The mass multiplication of *Trichoderma* on solid substrate promotes the synthesis of the desired enzymes which help in its biocontrol mechanism (Singh et. al. 2007).

7.1 Solid Sub-strate Fermentation

According to one school of thought, solid substrate fermentation (SSF) process is preferred over the other due to some inherent advantages under Indian conditions. A closer evaluation in recent years in several research centres throughout the world has revealed the enormous economical and practical advantages of SSF technology over submerged fermentation (SmF) technology. These include utilization of large number of agro-wastes as substrate for the en masse production of *Trichoderma.*

Advantages of SSF Technology

- A wide variety of matrices can be used as substrate
- Low capital cost
- Low energy expenditure
- Less expensive downstream processing
- Less water usage and low waste water out put
- Potential higher volumetric productivity
- Higher concentration of required products
- High reproductivity
- Lesser fermentation space
- Easier control of fermentation

7.2 Methodology for Preparation of Various Formulations of *Trichoderma* spp.

7.2.1 Talc-Based Formulation

Trichoderma culture biomass along with medium	-	1 lit
Talc (300 mesh, white colour)	-	2 kg
Carboxy methyl cellulose (CMC) or Arabic gum powder	-	10 g

Preparation Methodology : After growing *Trichoderma,* in any one of the above liquid media for obtaining biomass, it is mixed with talc powder in the ratio of 1:2 and dried to 8% moisture under shade. Carboxy methyl cellulose or Arabic gum powder should be added during mixing. After drying and breaking the clots, the formulation is packed in milky white polythene bags.

The technology of mass production of *T. viride* for seed treatment was developed by Jeyarajan et. al. (1994) and Ramakrishnan et. al. (1994) of Tamil Nadu Agricultural University (TNAU), Coimbatore. This formulation was supplied to farmers by the university to cover 20,000 ha annually during the past ten years. Among the various carriers tested to develop *Trichoderma* formulation, talc-based formulation retained its viability for 120 days (8×10^7cfu /g). Other carrier materials like lignite, peat and kaolin-based formulations had equal shelf-life. Hence, these carriers could be used for the formulation according to the material. The product stored for a period of 120 days gave equal control of root rot incidence in urdbean as that of the freshly prepared product.

Two industrial wastes namely precipitated silica and calcium silicate which are available in Madhya Pradesh were tested as carrier materials in the place of talc by Nakkeeran and Jeyarajan (1996). Precipitated silica is an article of commerce and mainly used wherever high absorbance and free flowing material is required. Calcium silicate is naturally occurring mineral used in ceramic and paint industry finely micronised (passing through 400 mesh) having high absorbance power with neutral pH and non-reactive at ordinary temperature. These granules gave a population of 99 and 104 $\times 10^6$ cfu/g respectively compared to 143 $\times 10^6$ cfu/g in talc substrate after 4 months of storage. These two substrates were also much cheaper than talc.

7.2.2 Vermiculite–Wheat bran Formulation (Lewis et. al.1991)

Ingredients

Vermiculite	-	100 g
Wheat bran	-	33 g
Wet fermentor biomass	-	20 g
0.05 N HCl	-	175 ml

Preparation Methodology

Trichoderma is multiplied in molasses-yeast medium for 10 days. Vermiculite and wheat bran are sterilized in an oven at 70 °C for 3 days. Then 20 gms of fermentor biomass and 0.05 N HCl are added, mixed well and dried in shade.

Longer shelf-life of *T. viride* was recorded with vermiculite-wheat bran acid fermentor biomass (VBA-FB) formulation stored in milky white bags (2.05 $\times 10^6$cfu/g). Studies on storage temperature revealed that 20-30 °C was optimum for storage the VBA-FB formulation in milky white bags. (Nakkeran and Devi., 1997).

7.2.3 Pesta Granules (Connick et. al. 1991)

(a) Wheat bran-based

Ingredients

Wheat flour	-	100 g
Fermentor biomass (FB)	-	52 ml
Sterile water	-	sufficient enough to form a dough

Preparation Methodology : 52 ml of FB is added to wheat flour (100g) and mixed by gloved hands to form cohesive dough. The dough is kneaded, pressed flat and folded by hand several times. Then 1 mm thick sheet (pesta) is prepared and air-dried till it breaks crisply. After drying, dough sheet was ground and passed through a 18 mesh (1.0 mm) sieve and granules were collected.

(b) Wheat flour – Kaolin (Prasad et. al. 1997)

Ingredients

Wheat flour	-	80 g
Kaolin	-	20 g
Fermentor biomass	-	52 ml

Preparation Methodology : 52 ml of FB is added to wheat flour (100g) and mixed by gloved hands to form a cohesive dough. The rest of the procedure is as described for pesta granules

(c) Wheat flour-Bentonite (Prasad et. al. 1997)

Ingredients

Wheat flour	-	80 g
Bentonite	-	20 g
Fermentor biomass	-	52 ml

Preparation Methodology : 2 ml of FB is added to wheat flour (100g) and mixed by gloved hands form cohesive dough. The rest of procedure is as described for pesta granules.

7.2.4. Alginate Prills (Fravel et. al. 1985)

Ingredients

Sodium Alginate	-	25 gm
Wheat flour	-	50 gm
Fermentor biomass	-	200 gm

Preparation Methodology : Sodium alginate is dissolved in one portion of distilled water (25g/750 ml) and food base is suspended in another portion (50g/250ml). These preparations are autoclaved and when cool are blended together with biomass. The mixture is added drop wise into $CaCl_2$ solution to form spherical beads, which are air-dried and stored at 5°C.

Numerous attempts have been made to control several soil borne pathogens by incorporating natural substrates colonized by antagonists of the pathogens into soil. The results of these attempts have varied according to substrate. Lowered disease severity, increased yield and a decreased pathogen population resulted from incorporation of different substrates colonized by *Trichoderma* spp.

in nurseries and experimental field plots. The importance of substrates in mass production of *Trichoderma* can not be undermined. The *Trichoderma* production process must also end up in biomass with excellent shelf-life even under adverse storage conditions. In several respects, the requirements for production of products for agricultural use are more difficult to meet than those required for pharmaceutical products. If agricultural materials are to be successful, they must not be very expensive, able to be produced in large quantities and maintain good viability without specialized storage systems.

Although there was a decline in the population of *Trichoderma* during storage in talc based formulation, a population of 230-242 × 10^6 cfu/g was observed at 20-30°C after 75 days of storage. *Trichoderma* can be safely stored as talc based formulations in milky white bags at room temperature upto 75 days.

Table 7.1 Comparison of substrate fermentation used for mass production.

Factors	Solid substrate fermentation	Liquid substrate fermentation
1. Substrates	Cheap substrates locally available	Soluble substrates (sugars)
2. Water	Limited consumption of water. No effluent	High volumes of water consumed and effluents discarded
3. Aeration	Easy aeration and high surface exchange air/substrate	Limitation of soluble oxygen. High level of air required
4. Mechanical agitation	Static conditions preferred	Required good homogenization
5. Energetic consideration	Low energy consuming	High energy consuming
6. Volume of equipment	Low volumes and low cost equipments	High volumes and high cost technology
7. Effluent and pollution	No effluents, less pollution	High volumes of polluting effluents
8. Concentration of products	Consisting of spores	Consisting of chlamydospores and basidiospores

7.3 Commercial Development of *Trichoderma*-based Biopesticide

In India, majority of farmers continue to express interest in biologically – based pest management strategies of all types as central components of an IPM approach. Such market realities promote the development of biocontrol products. Still, the path to developing and applying effective biocontrol methods is a long one, fraught with many difficulties. In addition to that, scientific, regulatory, business management and marketing issues must also be handled effectively for a biocontrol product to be successful in the market.

The following properties are essential for successful commercial production of *Trichoderma*-based biopesticide:

- Development of low-cost production, storage and distribution systems
- Ability to survive downstream processing, particularly drying, which is required to avoid contamination
- Stability and adequate shelf-life of the final product upon storage, preferably without refrigeration
- Ability to withstand environmental variations such as temperature, desiccation, relative humidity in order to survive and establish active populations in soil
- Consistent efficacy under varying agro-climatic conditions at commercial feasible states
- Integration of biocontrol into current agronomic practices
- Abundant and cost effective production of microbial propagules
- Optimization of culture conditions for producing high yield and high quality conidial biomass.

7.4 Ideal Characteristics of *Trichoderma* spp. for Commercial Exploitation

- Mycoparasitic activity against large number of economically important soil-borne plant pathogens.
- It should stimulate plant resistance and plant defense mechanisms.
- Ability to thrive over a wide range of stress conditions.

- Compatible with large number of chemical pesticides and weedicides.
- Enhance plant growth and productivity with efficient utilization; increase uptake and concentration of a variety of nutrients.
- Easy to mass multiply using solid substrate technology and formulate on suitable carriers.
- In addition to rhizosphere competence, it should also possess nematicidal properties.
- The final product should have more shelf-life.

Chapter – 8

Shelf-Life and Bio-efficacy of Biocontrol Formulations

A formulation can be defined as "the mixture of the candidate pesticide and material that affects its chemical, physical and biological properties". Formulation will be designed to retain the biological activity of the active ingredient, to prevent in store replication of any contaminant microorganism, largely by bacteria and fungi and to ease the handling of the product.

A formulation is required to present the product in a usable form and to optimize the efficacy, stability, safety and ease of application of the product. Thus, it deals with the methods of presenting the antagonistic microbes in the most effective form in terms of storage, application and ultimate biological activity. Developments of formulation involve combining biopesticides with carriers, diluents and surface active-agents, according to the predetermined formula to produce a final consumer product. It is very much essential to deliver the active ingredient at the point of use in form, to control the target organism effectively. A formulation should be cost-effective and convenient to apply. One of the purposes of formulation is the dilution of high concentration of the active ingredient down to a level, at which it may be toxic to the pest, but at the same time, it should not cause damage to desirable plants, natural enemies and animals.

Table 8.1 Differences between Pesticide and Bio-pesticide.

Pesticide	Bio-Pesticide
1. These are the organic products that do not contain living organisms.	1. These are the microbial products containing living organisms.
2. Lack of replication / multiplication in the environment.	2. Capable of replicating in the environment.
3. No need of multiplying to activate to control the target pathogen.	3. In order to control target pest, pathogen or weed it requires to get activated or multiplied after application.
4. Indiscriminate application leads to pest resurgence, environmental pollution.	4. Question of developing resistance by pathogens doest not arise and is eco-friendly in nature.

8.1 Types of Biocontrol Formulations

Different types of formulations are available such as granular products, water dispersible powder and emulsions.

8.1.1 Granules

Granular formulations are made from the clay minerals like attapulgite, montmorillonite and bentonite clays, granular diatomaceous earth and vermiculite and are being used as carrier materials. Granular formulations are usually applied for controlling soil-borne infections. These granules may also contain plant products. Some of the antagonists like *Trichoderma* are available in granular formulation.

8.1.2 Water Dispersible Powder

These are the most popular formulations used for the management of diseases. But the drawback associated is they are less efficient and environmentally hazardous, particularly when used for control of foliar diseases.

It consists of the microbial agent along with diluents such as talc, charcoal, attapulgite etc., and surface active agents like carboxy methyl cellulose (CMC), gelatin, gum etc., to ensure suspending and wetting properties. Some times nutrient base such as FYM powder, is added to support multiplication of bioagents after application. These formulations can be used for both seed treatment as well as foliar spray. The moisture content is one important characteristic that should be kept under control. It should not be more than 8% for fungi and 12% for bacteria (Anonymous, 1999).

For fungal antagonists like *Trichoderma,* the population should not be less than 10^6 cfu /g and of bacterial antagonists. Ex: *Pseudomonas, Bacillus* etc. or entomopathogens like *Beauveria, Metarhizium, Verticillium* the population should not be less than 10^8cfu/g of formulation. The shelf-life of WDP formulations of different biocontrol agents varies between 3 to 6 months which is estimated based on cfu/g of substrate.

8.1.3 Emulsions

In these formulations, microbial agents are suspended in a water immiscible solvent such as petroleum fraction (diesel, mineral oils) and vegetable oil (groundnut etc.) with the aid of a surface active agent. This can be dispersed in water to form a stable emulsion. Emulsifiable concentrates require a high concentration of an oil soluble emulsifying agent, to give instantaneous formation of a homogenous emulsion on dilution in water.

Such formulations of *Trichoderma, Pseudomonas* and *Beauveria* are now being used as foliar sprays. Liquid formulations of biocontrol agents are only recent introduction to Indian agriculture and its potential still to be explored.

The choice of the form of pesticidal product depends primarily on the type of pest involved, local weather conditions, the purpose for which the biopesticide is used.

It has been determined over the past decades that the carrier can play an active physical role in efficacy of the product. They can impart certain desirable physical properties to the product, thus making the choice of carrier extremely important. By definition, a carrier is an inert ingredient, without any chemical properties. The term is generally used to denote the inert ingredient used to dilute the biopesticide. The inert material that is used in the manufacture of dry biopesticide product does not have any inherent pesticidal activity. The commonly used carriers for the formulation of biocontrol agents include :

- Allumino silicate minerals like attapulgite
- Montmorillonite
- Bentonite
- Diatomaceous earth
- Vermiculite
- Kaolin
- Pyrophyllite

Inorganic compounds

- Talc
- Calcium and Magnesium carbonate
- Gypsum

Organic compounds like peat and lignite (charcoal) are considered as good carriers for biocontrol agents. The most widely used carrier for the biocontrol agents in India is talc. Type of carriers used in formulation may influence shelf-life. In one of the studies, shelf-life of *T .harzianum* was longer when kaolin and talc were used as carrier than bentonite (Prasad and Rangeshwaran, 2000). Addition of FYM powder to talc + CMC based formulation of *T. harzianum* enhanced temperature and UV radiation tolerance of the biocontrol agent.

8.2 Solid and Liquid Formulations

Majority of the biocontrol products available in the market are in the form of solid formulations i.e., granular or water dispersible powder. Development of oil-based formulations of biocontrol agents is recent area of research. The potential for development of mycopesticides has been boosted by the discovery that fungal conidia formulated in oils have shown greater infectivity than conventional water-based suspensions. Oil-based (paraffinic petroleum oil) formulations of *Trichoderma harzianum* (TH), *Pseudomonas fluorescens* (PsF) or their mixture (TH + PsF) showed better effect against brown spot of rice caused by *Cochliobolus miyabeanus* when used as foliar spray as compared to talc-based formulation of the same bioagents. Shelf-life of oil-based formulation was more than 10 months against 4 months for talc-based formulation at room temperature.

Advantages of Oil-based Formulations

Oils are non-phytotoxic, which substantially enhance the efficacy of hyper parasitic fungi. They are readily compatible with ultra low volume application techniques for spraying biopesticides at low humidity, being non-evaporative.

(i) The oils readily bind with the leaf surface and enhance penetration capacity of spray.

(ii) Uniformity of droplet size reduces the very fine droplets associated with the drift.

(iii) Spray oils coat the spray droplet to minimize evaporation problems. This allows fungal spores to germinate and become active even under low atmospheric humidity conditions.

(iv) Spray oils can be utilized to help overcome the problem of beaded droplets on waxy surfaces. Instead, the spray is evenly spread over the entire surface in a thin film.

(v) The problems associated with drizzle, wind and dew are reduced.

(vi) Dose requirement is much lower than WDP formulations.

Oil-based formulations are being developed with high population of *Trichoderma, Pseudomonas or Beauveria* ($>10^{10}$ cfu per ml). Some companies have recently developed a new concept in spray oils which are organic blend of emulsified, cold-pressed canola oil combined with highly penetrative Omega-3 fish oil. Fish oils are currently attracting considerable interest in relation to their apparent capacity to smother certain small insects and fungal diseases.

8.3 Shelf-Life and Bio-efficacy

Shelf-life of a biocontrol agent play a crucial role in storing a formulation. In general the antagonists multiplied in an organic food base has greater shelf-life than the inert or inorganic food bases. It has been realized that one of the major obstacles to commercialization of a biopesticide preparation is the loss of viability of biocontrol agents over time. Several attempts have been made to determine the viability of biocontrol agents in their preparations when stored at room temperature and refrigerator. Most of the results are variable and it is evident that shelf-life is also dependent upon species / isolate strain.

8.4 Development of Mixed Formulations of Biocontrol Agents

It is evident that most of the studies on biological control of plant diseases deal with single biocontrol agent as antagonist to a single pathogen. Single biocontrol agent is not likely to be active in all soil environments or against all pathogens that attack the host plant. Control of a wide spectrum of pathogens under a wide range of environmental conditions by applied antagonists largely remains an unfulfilled goal for biological control. Considering the fact that there is some degree of host-specificity in biocontrol agents even at sub-specific level, this may partially account for the reported in-consistent performance of biocontrol agent preparations.

A very important prerequisite for the possible use of antagonists in practical farming is that performance should be consistent under varying and unpredictable weather conditions. Because of the different sensitivities of biocontrol vis-a-vis environment conditions, the use of antagonists mixtures could contribute to consistency of performance of antagonists preparations. The mixture should be composed of organisms that could help each other to get a broader ecological range and a higher level of protection against different pathogens. If one organism fails due to any unfavourable condition, the other will be effective to overcome this problem. There are four main approaches:

- Select strains of biocontrol agent with wide host range
- Modify the genetics of the biocontrol agents that add mechanisms of disease suppression that are operable against more than one pathogen
- Alter the environment to favour the biocontrol agent and to disfavour competitive microflora
- Develop strain mixtures with superior biocontrol activity.

It is likely that most cases of naturally occurring biological control results from mixtures of antagonists, rather than from high population of a single antagonist. A combination of biocontrol agents is more likely to have a grater variety of traits responsible for suppression of one or more pathogens and also is likely to have these traits expressed over a wide range of environmental conditions. Combinations of biocontrol agents for plant diseases include mixtures of fungi and mixtures of fungi and bacteria. Most of these reports on mixtures of biocontrol agents showed that combining antagonists resulted in improved biocontrol. However, there are reports of combinations of biocontrol agents that do not result in improved suppression of disease compared with the individual antagonists (Dandurand and Knudsen, 1993). Incompatibility of co-inoculants can arise because biocontrol agents may also inhibit each other as well as the target pathogen or pathogens (Leeman et. al. 1996). Thus, an important prerequisite for successful development of strain mixtures appears to be the compatibility of the co-inoculated microorganisms (Baker 1990).

In acidic soils, *T. harzianum* (TH) exhibit better efficacy while *P. fluorescens* (PsF) is more effective against soil-borne diseases in neutral and alkaline soils. In a study, compatibility of 12 isolates of TH was tested against 41 strains of PsF under *in vitro* condition. In general, PsF suppressed the growth and sporulation of TH but 4 neutral combinations were identified. One of these combinations, TH strain PBAT-43 and PsF strain PBAP-27 was used to develop mixed formulation (Plant Biocontrol Agent-3), which was equally or more effective than individual formulations both under greenhouse and field conditions.

8.4.1 Strategies for Developing Mixtures of Biocontrol Agents

- Mixture of organisms with different plant colonization patterns.
- Mixtures of antagonists that control different pathogens.
- Mixtures of antagonists with different mechanisms of disease suppression.
- Mixtures of taxonomically different organisms.
- Mixtures of antagonists with different optimum temperature, pH or moisture conditions for rhizosphere / phyllosphere colonization.

Chapter – 9

Delivery of Biocontrol Agents

Any antagonist when introduced directly into soil faces stiff resistance/ competition from already existing resident soil microbes, will have advantage of being firmly established in soil by utilizing the available substrates. It is because of this reason, quite often an antagonist is applied through roots, seed or directly into the soil. It is unable to maintain high population in soil / rhizosphere beyond two to three weeks, especially in high organic content soils, resulting in loss of its efficacy against soil-borne plant pathogens. Thus, it results in inconsistent field performance by the introduced antagonist and is one of the major bottlenecks in biological control of plant pathogens. To overcome this problem, it is very much essential to have not only a qualitative biocontrol agent but also ensure its establishment in natural ecosystem where it can thrive well and compete with the pathogens. At this juncture, the role of delivery system is of paramount importance. But unfortunately the delivery system is one of the important areas of biocontrol research, which has been largely ignored. It is rather necessary to have an efficient, economic and ecologically viable mode of application of biocontrol agents in soil ecosystem. Like chemical pesticides, the biocontrol agents can be applied as:

- Seed treatment
- Seedling root treatment
- Soil treatment
- Foliar application

9.1 Seed Treatment

Seed treatment is one of the important methods of introducing biocontrol agents into soil-plant environment. The efficacy of seed treatment depends on the ability of the antagonist to multiply in the rhizosphere region. By this method, different vegetatively propagated materials like corms, bulbs, tubers and seed can be treated with antagonists to prevent seed, corm decay and seedling blight. Seed-borne inoculum can be effectively checked particularly externally seed-borne inoculum of diseases affecting above ground plant parts. Some of internally seed-borne diseases like loose smut of wheat is suppressed by seed treatment with *Trichoderma* (Selva Kumar et. al. 2000). This could be attributed to the action of metabolites produced by *Trichoderma* which may penetrate seeds to affect the growth of internally seed-borne inoculum. When an antagonist is applied through seeds, it will not only colonize spermoplane, shpermosphere, but also rhizoplane and rhizosphere. Different forms such as spore or cell suspension, dry powder are used to coat the seeds with potential antagonists. The propagules of biocontrol agents will germinate on the seed surface and colonize the roots of germinated seedlings (Tiwari, 1996). Based on seed size and formulation of antagonist, 3 to 10 g / kg of seed can be used for treating the seed (Mukhopadhyay et. al.1992).

Among the *Trichoderma* spp. *T. hamatum, T. harzianum, T. virens* and *T. viride* are effective seed protectants against *Pythium* spp. and *R. solani.* A large number of seed, seedling, root, stem, foliar and panicle diseases have been suppressed by seed treatment with antagonists.

As this technique of disease management represents a living system, the viability of biocontrol agent may pose a limitation to its commercialization. One approach for this might be supplying the coated seeds to farmers directly by the seed companies/ agencies. A considerable time gap between coating seeds by the seed supply agencies and sowing such seeds by farmers is bound to occur and it should be ensured that sufficient propagules remain viable on coated seeds at the time of sowing.

9.1.1 Seed Bio-priming

Seed bio-priming has potential advantages over simple control coating of seeds as it results in rapid and uniform seedling emergences. It involves slurry treatment of seeds with bio-agent in the presence of gum arabica, jaggery and / or FYM powder and subjected to incubation under high moisture for 24 to 48 h, results in improved performance of biocontrol agents against seed and soil-borne diseases. When the seeds are treated with bioagents like *Trichoderma* and *Pseudomonas,* it forms a thin layer around the seed surface, thus making the pathogens difficult to enter or invade. Bio-primed seeds could tolerate adverse soil conditions better.

Enhancement of *Trichoderma* population by 1000 times was observed under high moisture conditions. It is a superior method of control over simple seed treatment in a number of crops like rice, wheat, chickpea, tomato, brinjal, pea and bean as it resulted in better seedling emergence and plant health. The amount of inoculum required for bio-priming is less in comparison to simple seed treatment. It is usually 2 to 5 g bio-agent per kg seeds against 4 to 10 g bioagent per kg seeds for normal seed treatment.

This method is widely recommended as a general agricultural practice to be followed in most of the crops except soybean by the G.B. Pant

University of Agriculture and Technology, Pantnagar and State Government of Uttaranchal and Uttar Pradesh. Seed bio-priming may be avoided in crops like soybean which is sensitive to high moisture.

9.2 Seedling Root Treatment

This method involves treating the seedling roots with spore or cell suspensions of antagonists. It is practiced generally for vegetable crops, rice etc., where transplanting is practiced. Root dipping in antagonist suspension not only reduces disease severity but also enhances seedling growth in rice, tomato, brinjal, chilli and capsicum (Mishra et. al. 2001b). Application of *Trichoderma* spp. and /or *P. fluorescens* as root dip in tomato not only suppresses fungal diseases but also damage caused by root-knot nematode, *Meloidogyne* (Singh et. al. 2004).

9.3 Soil Treatment

Soil treatment is the most effective method of application of *Trichoderma* particularly for the management of soil-borne diseases. *Trichoderma* is capable of colonizing farmyards manure (FYM). For the purpose of soil application, the bio-agent is added to moist FYM and incubated for 10 to 15 days to allow colonization and colonized FYM is distributed in field followed by irrigation. *T. harzianum* was found ineffective when applied through rhizome, whereas colonized FYM could effectively suppress rhizome rot of ginger even in sick soil. Direct application of *Trichoderma* and / or *Pseudomonas* to soil by drenching is being widely used but its performance has been quite inconsistent.

Application of *Trichoderma harzianum* through colonized compost is the most effective delivery system as well as for *P. fluorescens* (Zaidi and Singh, 2004 b). In a number of other crops like tomato, brinjal, chickpea, lentil, pea, pigeon pea, capsicum, chilli, rajma, soybean, rice etc., soil application of *T. harzianum* and / or *P. fluorescens* through colonized compost was most effective in promoting plant growth and managing pre- and post-emergence damping – off, root rots and wilts. Other methods of application like seed treatment or root dip should be used only in combination with soil application of biocontrol agent colonized compost. When added through colonized FYM, these bioagents have the advantage of substrate colonization and are able to thrive in soil much better as compared to their application through any other means. *Trichoderma* and *Pseudomonas* could colonize and decompose poultry manure. *Trichoderma* also colonizes chicken manure whereas *P. fluorescens* not only multiplied very well in vermicompost but also significantly increased population and health of earthworms.

There are several reports on the application of biocontrol agents to the soil and other growing media either before or at the time of planting for control of a wide range of soil-borne fungal pathogens (Baby and Manibhushan Rao, 1996). Such applications are ideally suited for greenhouse and nursery but because of the bulk requirement, cost and problem of uniform distribution, feasibility of field application is less. Granular or pellet preparations have been used directly for soil application and they have provided effective control of diseases both under greenhouse and field conditions (Raghuchander et. al. 1993 ; Prasad and Rangeshwaran, 2000).

9.4 Foliar Application

Existence of epiphytic microflora on plant surface including leaves and flowers is a natural phenomenon. However, *Trichoderma* is primarily a soil fungus. Nevertheless, it has been used less frequently as a spray for the management of foliar diseases (Haware et. al. 1996; Mukherjee et. al. 1997). There are several reports showing effectiveness of biocontrol agents applied as foliar spray against different plant pathogens. However, success of any antagonist on leaf / sheath surface depends largely on its ability to colonize these surfaces. Even environmental factors like humidity, temperature and sun light affecting bio-efficacy of antagonists. Because of these reasons, foliar application is preferred during evening hours and it is more successful in rainy reason crops. In recent years, oil-based formulations of biocontrol agents like *Trichoderma* and *Pseudomonas* were found to be quite effective against diseases affecting foliage or blossoms (Singh et. al. 2004). These formulations are more effective than WDP formulations.

Chapter – 10

Biological Control of Soil-borne and Foliar Plant Pathogens

10.1 *Trichoderma* in the Management of Soil-borne Plant Pathogens

Management of soil-borne pathogens like *Rhizoctonia, Fusarium, Verticillium, Sclerotium, Pythium, Phytophthora* etc., using chemicals have been reported with limited success, since these pathogens can survive under a wide range of conditions such as high temperature, osmotic pressure etc. Apart from chemicals, other methods like growing of tolerant varieties, modified cultural practices, organic amendments etc., are in practice. Because of high-intensive agricultural practices, the traditional and age old eco-friendly disease management practices like sanitation, mixed cropping, adjusting time of sowing, crop rotation, summer ploughing, green manuring, composting etc., are not being practiced in Indian Agriculture. In spite of tremendous advancements made in the field of plant genetic engineering, the pace of development and durability of resistant varieties had been slow and unreliable.

Limitations of Chemical Control

- It is difficult to treat large volumes of soil/plant chemically to eliminate or reduce the pathogen population.
- Chances of developing resistance to fungicides, insecticides is high.
- Indiscriminate application leads to pollution of food and environment which further responsible for health hazards.
- High cost is involved.
- Efficacy is being affected by physico-chemical and biological factors.
- It also affects natural enemies, beneficial insects and predators present in the field.

Considering these limitations, there has been a growing awareness among farmers, researchers and policy-makers to develop such management practices, which alone or in combination with other practices could bring about a reasonably good degree of reduction of disease incidence and at the same time ensure sustainability of production, cost effectiveness and healthy ecosystem. Biological control is an important approach in this direction. Though biocontrol agents include different groups of organisms existing in an ecosystem, maximum emphasis is being laid on fungal and bacterial antagonists primarily because of ease with which they can be multiplied and formulated. In recent years, a critical review of literature reveals that control of plant diseases especially soil-borne diseases had been flooded with biocontrol agents like *Trichoderma* spp. and *Gliocladium* spp. Among the bacterial antagonists, the fluorescent pseudomonads are under intensive research because of their wide natural occurrence, biocontrol potential against fungal and nematode diseases as well as a host of defense-inducing and plant growth promoting activities. The biocontrol agents provide protection against plant diseases either by direct action against the pathogen by mycoparasitism or antibiosis and indirectly by competing with the pathogen for nutrients, oxygen or space. Therefore, biological control of soil-borne pathogens using antagonistic organisms has become very much essential in recent times.

Work on biological control in India and use of *Trichoderma* species gained momentum in 80's with efforts of Dept. of Plant Pathology from G-B. Pant University of Agriculture and Technology, Pantnagar, Uttaranchal and Tamil Nadu Agricultural University, Coimbatore. Presently several groups are actively engaged in work on biological control of plant diseases at ICAR institutes including the Project Directorate of Biological Control (PDBC), Bangalore, CSIR Institutes and State Agricultural Universities (SAU)

Though the researches in biological control of plant diseases started as early as in the 1930s through introduction of microorganisms, it took a backseat for the several decades. However, in the last 2-3 decades there has been an explosion of interest in biological disease control resulting in the availability of about 50 commercially formulated antagonist-based products (about 35 of fungal origin and 15 of bacterial origin) for plant disease management in several parts of the world (Mukherjee and Mukhopadhyay, 1997; Whipps 1997). In India, the pioneering work on the use of introduced antagonists for plant disease control had been initiated with works of R.S. Vasudeva and his associates. Their work on biocontrol of pigeon pea wilt by *Bacillus subtilis* and published a series of papers in Annals of Applied Biology in 1950-1965 are considered as a major landmark in Indian work on biocontrol.

However, the work on biocontrol in India really gained a momentum in the 1980s and the presidential address "Biological control of soil-borne plant pathogens by *Trichoderma* spp" delivered at the annual meeting of the Indian society of Mycology and Plant Pathology by Dr. Mukhopadhyay acted as a major stimulus to researches on plant disease biocontrol using *Trichoderma* spp.

Majority of the work done on plant disease biocontrol relates to soil-borne diseases using either bacterial or fungal antagonists. The early observation of Vasudeva and Roy (1950) explained the suppressive effects of *Bacillus subtilis,* an antagonistic bacterium resulted in low incidence of fusarial wilt of pigeon pea. He attributed this due to the release of antibiotic in the rhizosphere. Vasudeva et. al. (1952) developed a medium for the increased production of this antibiotic. Further the antibiotic was isolated and named as **Bulbiformin.** It was detected in the rhizosphere and under favourable conditions of antibiotic production, a marked reduction in the incidence of pigeonpea wilt was brought about by *B. subtilis*. Antibiotic production by *B. subtilis* can be markedly enhanced by amending soil with roots of certain leguminous crops, molasses and oil cakes. A reduction of 88% in the incidence of pigeon pea wilt was achieved under experimental conditions using autoclaved soil containing molasses, sweet clover roots and groundnut cake. When the seeds were bacterized with *B. subtilis* before sowing, the seedlings gained resistance to *Fusarium* infection and it was suggested that bulbiformin become systemic in the plant and provided a protective zone around the roots of pigeon pea seedlings.

The use of antagonistic fungi, especially *Trichoderma* and *Gliocladium* spp. has been more extensive than their bacterial counterparts. One of the earlier successes was reported by Padmanabhan and Alexander (1983). Approximately 3 lakhs of sugarcane seedlings were protected from root rot incidence by mass incorporation of *T. viride* in seedling flats.

Application of bioagents, *G. virens* and *T. longibrachiatum* through soil and seed suppressed *R. solani, S. rolfsii* and *P. aphanidermatum* in cotton, groundnut and tomato (Sreenivasa Prasad and Manibhushan Rao, 1990, 1993). A 30-95% reduction in the incidence of stem rot of groundnut was obtained by Asghari and Mayee (1991) through the application of *T. harzianum* and *T. longibrachiatum* under field conditions. Sheath blight of rice was controlled by integration of fungal antagonist and organic soil amendments.

A talc-based seed treatment formulation of *T. viride* has been extensively used by farmers of Tamil Nadu for protecting several crops from *Macrophomina phaseolina* infection (Jeyarajan et. al. 1993). Integration of carbendazim with *T. harzianum* and *T. viride* seed treatment resulted in reduced incidence of dry root rot and increased yield of soybean in farmers field (Vyas, 1994). In pot tests, seed pelleting with *T.*

harzianum, *Rhizobium* sp. and carbendazim was found to be highly effective in managing sclerotial root rot of groundnut.

Seed treatment with *Trichoderma* spp. and bacteria antagonist, *B. subtilis* controlled *M. phaseolina* in sesame under field conditions (Sankar and Jeyarajan 1996 a, b). A simple method of mass multiplication of *T. harzianum* was developed using coffee fruit skin composted with biogas slurry (Sawant and Sawant, 1996).

Karpagam (1995) and Mrinalini (1995) reported strain improvement of *Trichoderma* spp. using **protoplast fusion technique.**

A great deal of work has been done on biocontrol of soil-borne plant pathogens in India using *Trichoderma* and *Gliocladium* spp. at the GB Pant University of Agriculture and Technology, Uttaranchal. Research work on a wide range of pathogens viz., *S. rolfsii, R. solani, S. sclerotiorum, Fusarium* spp. and *P. aphanidermatum* using fungal antagonists *Trichoderma* spp. and *G. virens* had been carrying at Biocontrol Lab, Dept. of Plant Pathology, GB Pant University of Agriculture and Technology.

Soil incorporation of wheat bran preparations of *Trichoderma* in experimental and commercial nursery yielded good control of *Pythium* damping - off of many crops (Mukhopadhyay et. al. 1986; Mukherjee et. al. 1989, Sawant & Mukhopadhyay, 1990). Under field conditions, *T. harzianum* was highly effective in suppressing *S. rolfsii* in sugarbeet and wilt and root rot pathogens in chickpea. However, bulky nature of antagonist preparations restricts economical applications under field conditions. Therefore, thrust had been on the development of delivery system through seeds (Mukhopadhyay et. al. 1992).

A very high degree of protection against a range of pathogens was achieved by seed treatment with spores of *G. virens*. De and Mukhopadhyay, (1994) attributed this due to production of antifungal metabolites by the antagonist for biocontrol of P. *aphanidermatum*.

In recent years, a detailed study has been carried out with *G. virens*, an antagonist with a very broad spectrum of biocontrol activity. Mukherjee et. al. (1993) developed a technique for selective isolation of *G. virens* from natural soil. In this technique they have used *S. rolfsii* – colonized sorghum grains as a bait for antagonist were found superior to the conventional isolation techniques as selective media, because they eliminate the need of extensive screening antagonist isolates in a biocontrol programme. With reference to mode of action of antagonists, the principal mechanism involved in suppression of *S. rolfsii* and *R. solani* is due to parasitization of sclerotia. The ability of *G. virens* to parasitize *R. solani in situ* has been demonstrated (Mukherjee and Mukhopadhyay, 1995).

A simple protocol for obtaining stable mutants of *G. virens* through gamma irradiation has been developed by Mukherjee and Mukhopadhyay, 1993. One of the mutants produced (M+) which is non- conidiating, but produces profuse chlamydospores is an invaluable tool in studies on the role of chlamydospores in the survival and behaviour of this antagonist in soil, rhizosphere, fermentation media, and in storage. This mutant has been found to be as effective as wild type isolate in biocontrol potential against soil-borne pathogens in chickpea and lentil. *G. virens* has been tested extensively in farmer's fields and is being developed into a commercial formulation product.

Significant control of *Sclerotium* rot of groundnut by *T. harzianum* alone or in combination with fungicide was reported by Patibanda et al. (2002). Integration of *G. virens* and *T. harzianum* with carboxin as seed treatment was highly effective against seed and seedling rot of soybean under field conditions (Pant and Mukhopadhyay, 2001). Marked reduction of gladiolus corm rot or wilt was achieved by significant integrated corm treatment with *T. virens* (*G. virens*) and carboxin. Significant suppression of *Fusarium* population by *T. virens* in gladiolus field soil was achieved by Mishra et. al. (2003). Central Institute of Medicinal and Aromatic plants (CIMAP), Lucknow and National Botanical Research Institute (NBRI), Lucknow were able to patent *Trichoderma* product / process.

Apart from *Trichoderma* and *Gliocladium* spp. some other fungal antagonists like *Aspergillus terreus* (Singh and Dwivedi, 1992), *A. nidulans* (Upadhyay and Rai 1992), *A. niger* (Chattopadhyay and Sen 1996) and vesicular arbuscular mycorrhiza have also been successfully used for disease biocontrol under experimental conditions. **Kalisena,** a biopesticide based on *A. niger* developed by Dr. Bineeta sen, has been commercialized for large-scale field application. Other bioagents that are reported so far for biological management of plant diseases are as follows:

10.2 *Sporidesmium sclerotivorum*

In nature, *Sporidesmium sclerotivorum* behaves as an obligate parasite on sclerotia of *Sclerotinia sclerotiorum, S. minor, S. trifolicum cepivorum* and *B. cinerea* and it appears to be a nearly perfect biological control agent (Adams, 1990). One of the interesting features of *S. sclerotivorum* mycoparasitism is that their glucanase activity results in the degradation of glucan to glucose which the mycoparasite is able to utilize.

10.3 *Coniothyrium minitans*

This mycoparasite appears to have great potential for biological control of a number of diseases caused by *Sclerotinia* spp. and *S. cepivorum.* Field control of *S. sclerotiorum* in sunflower and glass house control of *S. sclerotiorum* in sunflower in lettuce have been achieved following soil incorporation of solid substrate formulations before planting. It is generally considered to be a parasite of ascomycetous fungi. There are many examples, mostly from *in vitro* studies that it can attack the hyphae of a range of fungi (Whipps and Gerlagh, 1992).

10.4 *Fusarium* spp.

The saprophytic *Fusarium* spp. isolated from *Fusarium* suppressive soils have been extensively and successfully exploited for their activity against various formae speciales of *F. oxysporum*, incitant of vascular wilt of various economically important crops. The high level of disease control provided coupled with the consistent of results obtained made saprophytic *Fusarium* spp. as a good biocontrol agent and holding promise for practical application on crops like basil, watermelon and cucumber (Gullino et. al. 1995; Larkin et. al. 1996). Induced resistance has been suggested as another mechanism of disease reduction in chickpea against wilt (Hervas et. al. 1995).

10.5 *Pythium oligandrum*

The ability of *Pythium oligandrum* to rapidly parasitize or lyse other fungi viz., *Pythium ultimum, Fusarium oxysporum, F. culmorum, B. cinerea* and *R. solani* led to the study of its potential as seed applied biocontrol agent. Oospore formulations of *P.oligandrum* has recently been used for biological control of *Pythium* damping - off in cress (Mc Quilken et. al. 1992).

10.6 *Talaromyces flavus*

Talaromyces flavus is considered to be a potential biocontrol agent of several plant pathogens. It is a mycoparasite of *R. solani, S. sclerotiorum* and *Verticillium dahliae.* Mc Laren et. al. (1994) obtained biocontrol of *Sclerotinia* wilt of sunflower with *T. flavus.* Fravel et. al. (1986) showed that *T. flavus* suppressed *Verticillium* wilt of potato when applied to soil as pelletted formulation. Results in the same field the next year indicated that *T. flavus,* had the potential to carry over as a biocontrol agent. Glucose oxidase is the protein secreted in the culture filtrate of *T. flavus* which is responsible for the inhibition of sclerotial germination of V. dahliae.

10.7 *Ampelomyces quisqualis*

It is a parasite of many powdery mildew pathogens including *Sphaerotheca fuliginea* and *Erysiphae cichoracearum* that attack cucumbers and *Uncinula necator* that attack grapes. The mechanism of action involves that the antagonist penetrates from cell to cell through septal pores of the powdery mildew fungi and continues to grow during gradual degeneration of the infected cells. This mycoparasite is also reported to parasitize the cleistothecia of *U. necator* (Falk et. al. 1995). Application of antagonist immediately after appearance of *S .fuliginea* colonies on cucumber foliage, efficiently suppressed mildew and increased yields of marketable fruits (Sundheim, 1986). In a recent study, *A. quisqualis* was grown on nutrient treated cotton twine suspended over grapevines. The grapevine powdery mildew caused by *U. necator* was suppressed by the pycnidiospores of mycoparasite throughout the rainy season.

10.8 Yeasts

Saprophytic yeasts have been explored as biocontrol agents for foliar and post harvest diseases. They are ideally suitable for the management of post-harvest diseases because,

(i) they can colonize the wound surface under dry conditions,

(ii) produce extracellular polysaccharides that enhance their survivability,

(iii) restrict the colonization sites and of germination cause to fungal propagules,

(iv) rapidly utilize the available nutrients and proliferate, and

(v) they are minimally impacted by pesticides.

The osmophillic yeast *Debaromyces hansenii* and the common yeast, *Candida oleophila* have shown great promise in bioprotection of fruits from post-harvest decay by fungi.

Isolates of yeasts viz., *Rhodotorula lutinis* and *Cryptococcus albidus* were found to be effective against grey mould caused by *Botrytis cinerea* in bean and tomato plants (Elad et. al. 1996). The mechanism of action involves competition for nutrients with the germinating conidia, and induced resistance. Certain yeasts like *Pichia quilliermondi* and *Tilletiopsis* spp. are reported to be mycoparasites on some pathogenic fungi.

10.9 Biological Control of Foliar Plant Pathogens

During the last decade, lot of emphasis is being laid on biological control of foliar plant pathogens but most of them are unsuccessful attempts. Many pathogens were tested under *in vitro* or under controlled conditions, these could not be successfully transferred to field condition because of fluctuating environment. However, research efforts are continued because of,

- increased failure in current pesticides lead to environment pollution,
- development of resistance to fungicides by pathogens, and
- contamination of food

New vistas are opened up due to genetic engineering by developing improved strains. Commercial formulations of *Trichoderma* are available in countries like USA and Israel for control of some foliar diseases. For foliar diseases like powdery mildew and downy mildew of various crops, turf grass diseases like brown patch caused by *R. solani,* damping – off (*Pythium* spp.) and *Botrytis cinerea*, the bioagent *T. harzianum* strain T-22 available under trade names **TOP shield, Bio-Trek 22 G, T22 Hopper Box** in USA was found to be effective.

In USA, a commercial product 'Grey gold' was developed for control of B. cinerea, which consists of a mixture of fungi *T. hamatum* and *Rhodotorula glutinis* and the bacterium *Bacillus megaterium.* For control of grey rot caused by *B. cinerea* and other foliar diseases in various crops, a commercial product 'Trichodex' based on *T. harzianum* strain 39 is available in Israel. In India no commercial products are reported for foliar disease control at present but a few reports have shown the possibility of using bioagents for management of foliar disease.

Examples of mycoparasites occur on foliar plant pathogens

Mycoparasite	Foliar diseases
Darluca filum	Several rust species
Tuberculina maxima	*Cronartium ribicola*
Ampelomyces quisqualis	Powdery mildews
Tilletiopsis sp.	Powdery mildew of cucumber
Nectria inventa	
Gonatobotrys simplex	*Alternaria* leaf spot

10.9.1 *Trichoderma* in Foliar Disease Management

Mukherjee and Haware (1993) isolated a *T. viride* isolate form chickpea rhizosphere which was found to be highly effective against the grey mould, *B. cinerea* of chickpea both under greenhouse and field conditions

(Mukherjee et al. 1995). Field trials were conducted to test the efficacy of integrated control methods for the control of grey mould of chickpea caused by *B. cinerea*. The plots with treatments viz., chickpea detopped at 60 DAS + *T. viride* + Vinclozolin sprays showed the lowest disease severity and highest grain yield. The combinations of antagonist and fungicide were superior to fungicide or antagonist alone (Abha et. al. 1999).

Field experiments were conducted to study the control of early blight of tomatoes (cv. Pusa Ruby) caused by *Alternaria solani*. Of the *T. viride* treatments, seed treatment combined with soil application was the most effective, recording the lowest disease incidence (Sawant et. al. 1999). Studies on the efficacy of chemical and biological treatments for the control of covered smut of barley caused by *Ustilago hordei* (*U. segetum*) showed that infection of smut fungus could be eradicated completely in fungicidal treatments as well as the *T. viride* treatments (Singh, 1999).

In controlling early and late leaf spot of groundnut caused by *Cercospora arachidicola* (*Mycosphaerella arachidis*) and *Phaeoisariopsis personata* (*M. berkeleyi*), respectively and rust (*Puccinia arachidis)* in groundnut and alternaria blight (*A. helianthi*) in sunflower, *T. viride* was found to be equally effective as that of fungicides (Mathivanan et. al. 2000). The work carried out at PDBC, Bangalore on biological control of botrytis grey rot of rose resulted in identification of potential *Trichoderma* isolates (Prasad et. al. 1998).

Fungal biological control agents were used either alone or in combination with fungicides to manage black spot of rose caused by *Diplocarpon rosae* under greenhouse conditions. The disease incidence was significantly low in biocontrol agent and fungicide treatments alone or in combination treatments compared to pathogen check at all observation dates (disease ratings from 0.33 to 3.33).

Chapter – 11

Bioagents in Management of Post-Harvest Diseases of Fruits

Losses from post-harvest fruit diseases range from 1 to more than 20 per cent in United States depending on the commodity and can be as high as 50% in developing countries. Application of fungicides to fruits after harvest to reduce decay has been increasingly curtailed by the development of pathogen resistance to many key fungicides, the lack of replacement fungicides, negative public perception regarding the safety of pesticides and consequent restriction of fungicide use. Biological control of post-harvest diseases has emerged as an effective alternative. The first commercial biocontrol products registered in United States by the US Environmental Protection Agency are sold as **'Biosave 100'** and **110** and **'Aspire'**. In other countries like South Africa, biocontrol products are registered and sold as **'Avogreen'** and **'Yieldplus'**.

11.1 Uniqueness of Post-Harvest System

Wounds made during harvesting and fruit handling can be protected from wound invading pathogens with single post-harvest (PH) application of the antagonist directly to wounds. The levels of decay acceptable in PH systems are generally below 5%, a standard that is often achieved with single application of (Biocontrol Agents) BCAs. Stress tolerant and competitive species would require different biocontrol strategies. Controlling rots resulting from preharvest latent infections by PH treatments is difficult but successful control of latent infections by PH applications have been reported.

E.g. *Bacillus spp.* are dominant colonizers of phylloplane and fructoplane which control latent infections caused by *Colletotrichum spp.* on mango and avocado.

In biocontrol of anthracnose of avocado, *Bacillus subtilis* applied as a dip and wax formulation removes nutrient from the immediate surrounding of the appresorium of *Colletotrichum gleosporioides* to ensure maintenance of the dormancy. The main strategy used to suppress PH fruit decay is the post harvest application of antagonists to prevent pathogens from infecting fruit wounds after harvest. Post harvest decay can also be suppressed by field application of BCAs. Eg: Field sprays against anthrancose on mango and avocado and *Pezicula malicorticis* on apples.

To control post harvest decays originating from latent infection in orchard, the application of antagonists must be repeated. e.g., Three sprays of *B. subtilis* and three sprays of copper controlled anthracnose of avocado. Continuous application of *B. subtilis* over a 4-year period resulted in gradual buildup of the antagonists, reduction of the level of pathogen inoculum and more effective control.

11.2 Sources and Isolation of the Antagonists

The fruit surface is an excellent source of naturally occurring antagonists against post harvest fruit decay. Isolation of the antagonists can be improved by using fruits from unmanaged orchards where natural populations have not been disturbed by chemical usage and the pool of potential antagonists is greater than in a chemically managed orchard. A variety of enrichment procedures are used for favorable isolation of microorganisms growing efficiently on the substrate.

- Isolation from natural cracks on the fruit surface.
- Agar plates containing apple juice that were seeded with fruit washings.
- Fruit wounds treated with washings and incubated for several days.
- Freshly made wounds on apples in the orchard that were exposed to colonization by fruit associated microbiota from 1 to 4 weeks before harvest.

The fructoplane and phylloplane provide the most abundant and most desirable source for isolating antagonists against post harvest fruit pathogens. Screening collections of yeast / starter cultures used in food industry may also yield effective antagonists. Soil also may be an abundant and diverse source of antagonists.

E.g. *B. subtilis* B_3 a contaminant, found during routine isolation of fungi from peach roots and identified for its inhibition of *M. fructicola* on agar plates is a very effective strain for biocontrol of brown rot of peach.

11.3 Mechanisms of Biocontrol

The different mechanisms reported for the control of post-harvest diseases are antibiosis, production of lytic enzymes, parasitism, induction of resistance and competition for limiting nutrients and space.

11.3.1 Antibiosis

Antibiotics, Iturin is produced by *B. subtilis* B_3 and Pyrrolnitrin is produced by P*seudomonas cepacia* LT-4-12w, reduced *in vitro* growth and conidia germination of stone fruit pathogen *M. fructicola* and pome fruit pathogens *P. expansum* and *B. cinerea.*

Strain LT-4-12w of P*seudomonas cepacia* also controlled various decays on citrus and stone fruits. *Pseudomonas syringae* strains Esc-10 and Esc-11 used in Biosave products produce **Syringomycin E**. which is inhibitory to a variety of fungi such as green mold of lemons.

Yeast, *Pichia anomala* strain K effective in the control of greymold of apple increased production of exo-β-1, 3-glucanases threefold in the presence of cell wall preparations of *B. cinerea* in apple wounds, reducing lesion size by more than half compared to antagonist alone. Higher β-1, 3-glucanases and chitinases activity was detected in fruit wounds treated with strains of an antagonist *Aureobasidium pullulans* in controlling various decays on apple, table fruit and other fruits. *A. pullulans* also produce antibiotics, **Aureobasidins**. *Erwinia herbicola* strains B_{66} and B_{90} controlling blue and grey mold of apple, inhibited conidia germination and lysed germtubes of *B. cinerea* and *P. expansum* in diluted apple juice. The yeast *Candida famata* reduced green mold decay on oranges and increased the phytoalexins, **scoparone** and **scopoletin** 12 folds in fruit wounds after 4 days when inoculated alone.

11.3.2 Competition for Space and Nutrients

Attachment of antagonists to pathogen hyphae has been suggested as an important factor in competition for nutrients in the case of *Enterobacter cloacae* for control of *Rhizopus stolonifer* on peach, *Pichia guilliermondii* for control of *Penicillium italicum* on citrus and *P.guilliermondii* for control of *Botrytis cinerea* on apple. *Enterobacter cloacae* inhibit conidial germination and decay, Production of ammonia has been reported as a mechanism of biocontrol.

Reports on the mechanisms to improve biocontrol of plant disease suggest that competition for nutrients and space plays a major role in most cases. Rapid colonization of fruit wounds by the antagonists is critical for decay control and manipulations leading to improved colonization and enhanced biocontrol. Within microbial communities, interactions are density-dependent and more than one type of interaction can occur at any one time depending on the growth phase of different microorganisms, population density and species diversity. Three types of interactions reported are **competition for nutrients**, **competition for space** and

inhibition by secondary metabolites. They were observed with preharvest sprays of *B. subtilis* to control *C. gleosporioides* on avocado.

11.4 Main Approaches for Post-Harvest Disease Control

The following strategies are adopted for the management of post-harvest diseases effectively.

- Manipulation of environment
- Use of mixed cultures of antagonists
- Physiological and genetic manipulation of antagonists
- Combining field and post harvest applications
- Manipulation of formulations
- Integration with other methods

11.4.1 Manipulation of Environment

It involves manipulation of the physical and chemical environment to the advantage of antagonists in storage, but at the same time fruit quality must be maintained. Fruits are often treated and or handled in a water suspension before, during and after storage, which provides an excellent opportunity to modify the environment. Nitrogen is likely to be a limiting nutrient in the carbon-rich environment of apple and pear wounds. Addition of **L-aspargine** and **L-proline** enhanced population of the antagonist *P. syringae* more than 10 fold during the critical first 24 hrs at room temperature resulting in reduction of blue mold decay from 50% to 0%. Addition of a nutrient analog 2-deoxy-D-glucose which is taken up by the pathogen is not metabolized, thus inhibiting pathogen growth giving advantage to antagonists *P. syringae, S. roseus* and C*andida saitoana,* thus improving biocontrol of decay of apple and citrus. Addition of siderophores may reduce apple decay by sequestering iron required for germination of some post-harvest pathogens. If nutrients limiting growth of pathogens or antagonists are known, then limiting nutrients / substances could be manipulated to stimulate antagonist population and or biocontrol mechanisms.

11.4.2 Applying Mixed Cultures

Mixtures of compatible strains may be needed to provide broad spectrum of activity. The antagonistic action result not from an activity of one species but rather from the action of community of microorganisms that suppress a target pathogen through different mechanisms of action. The application of antagonist mixtures has reduced variability and improved efficacy of biocontrol in many systems.

11.4.3 Physiological and Genetic Manipulations

Physiological manipulations of antagonists has been focused on improving their ecological fitness, which is particularly important in orchard applications. Populations of low water activity (a_w) tolerant cells of the antagonist *Candida sake* obtained from growth on media modified for low water activity and applied to apples 2 days prior to harvest, enhanced the biocontrol activity towards blue mold. Physiological manipulation may also be used to enhance mechanisms of biocontrol. This has been demonstrated for soil antagonists, where addition of zinc increased production of antibiotic **'phenazine'**, which is a mechanism by which *P. fluorescens* suppresses take-all of wheat caused by *Gaeumannomyces graminis var. tritici.* Iron sequestration by a siderophore, **'rhodotorulic acid'**, produced by the yeast *Rhodotorula glutinis* is responsible for biocontrol of *P. expansum* on apples. Addition of siderophore to antagonist suspension increased biocontrol activity. Antagonists are genetically modified to improve biocontrol of plant diseases. Efforts are focused on development of efficient transformation procedures for bacterial and yeast antagonists, and inserting genes for tracking the antagonist in the environment rather than enhancing biocontrol. Attempts are being made to overexpress genes involved in biocontrol. e.g., Lytic enzymes or engineering strains with desired biocontrol traits may yield positive results.

11.4.4 Pre-Harvest Applications

There are two distinct approaches in applying biocontrol agents in the field to control post-harvest decays of fruits.

- Antagonists developed for preharvest applications are applied just before harvest. The intent is to precolonize the fruit surface with an antagonist immediately before harvest, so that wounds inflicted during harvest can be colonized by the antagonist prior to colonization by a pathogen.
- Antagonists selected for field application are applied throughout most of the fruit development, to reduce latent infections that can originate as early as bloom time and cause fruit decay after harvest.

E.g. the antagonistic yeast *Cryptococcus infirmo- miniatus, C. laurentii* and *R .glutinis* applied to d'Anjou and Bosc pears in the field three weeks before harvest reduced grey mold on Bosc pear from 13% to as little as 4% and on d'Anjou pears from 7% to as little as 1%.

Populations of antagonists such as *A. pullulans, R. glutinis* and *B. subtilis* increased after application to apples in the orchard and maintained at relatively high densities on fruit in cold storage except for *B. subtilis* which declined. Applying mixtures of antagonists in the orchard may be useful for controlling post-harvest decays, especially if applied in the early growing season, which may reduce latent infection caused by P*ezicula spp. Candida sake* CPA-1 reduced blue mold by 50% on wounded apples inoculated with the antagonist 2 days prior to harvest. Field infections of *Botrytis* are effectively controlled by preharvest sprays of *Gliocladium roseum, Trichoderma harzianum, B. subtilis, B. licheniformis,* yeast and other bacteria.

11.4.5 Manipulations of Formulations

Formulations have profound effect on biocontrol agents and products, including shelf-life, ability to grow and survive after application, effectiveness in disease control, ease of operation and application, and cost. Certain freeze-drying protective agents and dehydration media enhance the viability of the antagonist P*antoea agglomerans* strain CPA-2 effective against blue mold and grey mold of pome fruits. The greatest protection from freeze-drying injury to the antagonist was provided by 5% trehalose, with the survival of over 60% of cells, whereas 10% skim milk was best dehydration medium resulting in a 100% recovery of freeze-dried bacterium. The addition of xanthan gum to *A. pullulans* L_{47} applied to strawberries in field from bloom to fruit at the green stage improved survival of the antagonist and increased biocontrol of storage rot by *B. cinerea*.

Formulations can influence the survival and activity of BCAs on fruit surface and in wounds. In some spore forming bacteria such as *Bacillus* spp. breaking the endogenous dormancy of the formulated product may be a factor in speed of action which is crucial in post-harvest biocontrol. A shelf-life of less than 6 months requires direct order service whereas 0.5 to 2 years is adequate for conventional off-shelf sales. The biocontrol products Biosave 100 and 110 have been formulated as frozen pellets and as a wettable powder.

11.4.6 Integration with other Methods

Applying a 2% solution of $CaCl_2$ together with the yeast antagonist, C*andida* spp. enhanced biocontrol of grey and green mold on apples. Application of 68 mM $CaCl_2$ to grapefruit reduced the incidence of green mold decay by 43% and in combination with an antagonist P. *guilliermondii* strain US-7 by 97%. Prestorage hot air treatments (38 °C for 4 days) of apples reduced or eliminated blue mold decay caused by *P .expansum* and grey mold decay. Hot water (45-56°C) dip treatments are used commercially in citrus and mango packing houses in South Africa to reduce post-harvest decays. The efficacy of control could be enhanced by the addition of *Bacillus spp.* to the hot water dips. Chitosan and its derivatives, including glycochitosan, were reported to inhibit fungal growth and to induce host-defense responses in plants and harvest commodities. Combining 0.2% glycochitosan with the antagonist *C. saitoana* is effective in controlling green mold of oranges and lemons caused by *P. digitatum.* A fruit coating containing sodium salts of carboxy-methyl-cellulose, sucrose esters of fatty acids and mixed sucro-glycerides and soap, commercialized under the name **TAL Pro-long** reduced the spread of a range of post harvest decays of pome fruits. Irradiation of root vegetables such as carrots or sweet potatoes with UV-C (wavelength <280nm) induced resistance and reduced post-harvest decays. Ethanol at 10% in combination with ethanol resistant *S. cerevisiae* strains 1440 and 1749 isolated from wine and ensile corns , reduced the incidence of grey mold decay on apples from more than 90% to close to 0%.

11.5 Case Studies

The expression of an antifungal peptide in *Saccharomyces* to enhance its biocontrol activity against post-harvest disease of tomato caused by *Colletotrichum coccodes* was studied by Richard W. Jones and Dov Prusky in 2002. In this experiment, a **Cercopin A**-based peptide capable of inhibiting germination of *Colletotrichum* was cloned into *Saccharomyces cerevisiae.* These yeast transformants inhibited decay development caused by *C. coccodes* in tomato. Expression of antifungal peptide in yeast therefore represents a new approach for the biological control of post - harvest disease. This antifungal peptide was synthesized by Genosys (Genosys, The Woodland's Tx).

11.5.1 Antifungal Activity of Yeasts

The biocontrol efficiency of antagonist yeasts of gray mold and blue mold on Apples and Pears in controlled atmospheres was studied by Shiping Tian *et. al.* (2002). Biocontrol capability of the yeasts *Trichosporon* sp. and *Cryptococcus albicus* against *Botrytis cinerea* and *P. expansum* was evaluated in apple (cv. Golden Delicious) and pear (cv Tingbai) fruits at 1°C in air and under controlled atmospheres (3% O_2 + 3% CO_2 (or) 3% O_2 + 8% CO_2). The antagonists *Trichosporon* sp. and *C. albicus* were more effective for control of apple decay in CA with low O_2 + high CO_2 than in air. *Trichosporon* sp. controlled gray mold and blue mold of apple fruits more effectively than *C. albicus.* Apple fruits treated with *Trichosporon* sp. and *C. albicus* has a lower incidence of gray mold rot than blue mold rot in the same storage conditions. Biocontrol efficacy of the yeasts for controlling gray mold and blue mold was better in apples than in pears.

The post-harvest biological control of *Rhizopus* rot of nectarine fruits by *Pichia membranefaciens was reported by* Fan Oing and Tian Shipping

(2000). A new yeast antagonist, *Pichia membranefaciens* isolated from wounds of peach fruit was evaluated for its biocontrol ability against *Rhizopus stolonifer* on nectarine fruits at different temperatures (25, 15, 3 °C) and with other treatments. In the experiment, it was observed that *P. membranefaciens* at 5 × 10^8 CFU/ml of washed-cell suspension completely inhibited *Rhizopus* rot at 25, 15 and 3 °C. A culture filtrate of the yeast antagonist (treatment B) failed to provide any protection against *Rhizopus* rot in nectarine fruits. In the same experiment they also studied the effect of wounding and applying *P. membranefaciens* prior to the inoculation of nectarine fruits with *Rhizopus stolonifer* for control of rot. Rapid colonization of yeast in wounds was observed during the first 48 h at 25°C and 15°C. *P. membranefaciens* at 5 × 10^8 CFU/ml was effective when applied 0 to 72 h before the pathogen, while at 1 × 10^8 CFU/ml, its efficacy was significantly reduced when yeast was applied simultaneously with the pathogen with disease incidence of 60%.

Control of post-harvest decay in pear by four laboratory grown yeasts and two registered biocontrol products was demonstrated by David and Robert A. Spotts. Laboratory grown yeasts include – *Cryptococcus infirmo-miniatus, Rhodotorula glutinis* and 2 strains of *C. laurentii.* Two registered biocontrol products include **Aspire** containing *Candida oleophila* and **Biosave-11** (NW Biosave-110) containing the bacterium *Pseudomonas syringae.* In this experiment the laboratory grown yeasts and registered biocontrol products were used alone and also in combination with 100 ppm thiobendazole. In addition, Aspire with 2% $CaCl_2$ to test their ability to control blue mold pathogen, *Penicillium expansum (*both thiobendazole sensitive and TBZ – resistant strains). Among all the treatments, the treatments with laboratory grown yeasts and registered biocontrol products in combination with TBZ 100 ppm showed complete inhibition in blue mold lesion development in TBZ-sensitive *Penicillium* strains to about 89% in TBZ-resistant *Penicillium* strains.

Attachment of antagonists to pathogen hyphae has been suggested as an important factor in competition for nutrients in the case of *Enterobacter cloacae* for control of *Rhizopus stolonifer* on peach, *Pichia guilliermondii* for control of *Penicillium italicum* on citrus and *P. guilliermondii* for control of *Botrytis cinerea* on apple. *Enterobacter cloacae* inhibit conidial germination and decay, **production of ammonia** has also been reported as a mechanism of biocontrol.

Among the individual treatment with laboratory grown yeasts and registered products, *Cryptococcus laurentii* strain 87-108 was more effective in controlling blue mold pathogen than all other treatments.

Post-harvest biocontrol of green and blue mold and sour rot of citrus fruit by *Debaromyces hansenii* was studied by Edo Chalutz and Charles

(1990). The yeast, *Debaromyces hansenii* isolated from the surface of lemon fruit, inhibited incidence of green mold, blue mold and sour rot of several citrus fruit cultivars. Among six cultivars of citrus, *D. hansenii* was effective in controlling green mold than other two rots that too in pumelo followed by grape fruit.

The biocontrol efficacy of *Candida saitoana* and glycochitosan for the control of post-harvest diseases of apple and citrus fruits was studied by El-Ghouth *et al.* (1999). Recently a biocontrol product called '**bioactive active coating**' consisting of a unique combination of an antagonistic yeast with chemically modified chitosan was developed. The bioactive coating was more effective in controlling decays (blue mold and green mold) of several cultivars of apples (Red Delicious, Rome, Golden Delicious and Empire) than either *C. saitoana* or 0.2% glycochitosan. Depending upon the apple cultivar used, the bioactive coating was comparable or superior to thiobendazole in reducing decay.

Apples

On cultivars **Golden Delicious** and **Rome**, the bioactive coating reduced decay significantly more than *C. saitoana* or 2% glycochitosan alone. In all apple cultivars tested, except Rome, the bioactive coating gave significantly better control than thiobendazole.

Citrus

Decay of oranges and lemons was higher on late season than on early season fruit. On early season oranges and lemons, the bioactive coating and imazalil were the most effective treatments than *C. saitoana* and 0.2% glycochitosan.

Chapter – 12

Phytoalexins in Plant Disease Control

Phytoalexins are the low molecular weight fungitoxic substances produced by plants as a result of infection by the pathogen or due to abiotic stresses like mechanical or chemical injuries. It is defined as these are the 'antibiotics produced in plant pathogen interaction or as a response to injury or other physiological stimuli'. Phytoalexins do not exist in the plant or pathogen systems. But they are formed as a result of host-parasite interaction. E.g. Potato tubers produce phytoalexin, **'Rishitin'** only after infection with *Phytophthora infestans*. Phytoalexins are not only toxic to fungi, but they are also toxic to bacteria, nematodes and other organisms.

12.1 Distribution and Biosynthesis of Phytoalexins

12.1.1 Distribution

More than 350 phytoalexins have been chemically characterized from approximately 30 plant families. The greatest number 130 have been characterized from the Leguminosae. Most of the phytoalexins have been isolated from dicotyledons but they have also been isolated from monocotyledons including rice, corn, sorghum, wheat, barley, onions, and lilies. The phenyl propanoid phytoalexins are distributed among families as diverse as the Leguminosae, Solanaceae, Convolvulaceae, Umbelliferae and Gramineae. In Leguminosae three groups of phytoalexin compounds have been demonstrated and characterized. These are Pterocarpons, Isoflavans and Isofavanones. In Solanaceae, only terpenoid phytoalexins have been characterized. Within a family, a level of specificity also is evident, e.g., the non-sesquiterpenoid phytoalexin, **rishitin** is found in potato and tobacco but not in pepper, whereas the sesquiterpenoid phytoalexin, **capsidiol** is found in tobacco and pepper but not in potato.

12.1.2 Biosynthesis of Phytoalexins

The phytoalexin synthesis is parallel to the mechanisms of defense through production of antibodies in the animal system. Phytoalexin production and also accumulation occur in healthy cells surrounding infected or wounded cells and are stimulated by some substances (elicitors) produced and released by these infected cells. The major phytoalexin biosynthetic pathways are **shikimate**, **Acetate-malonate** and **Acetate-mevalonate** pathways and are produced from one of the 3 pathways or a combination of 2 or 3 of the pathways.

- Chlorogenic acid phytoalexin synthesized by shikimate path way.
- Ipomeamarone phytoalexin synthesized by Acetate-mevalonate pathway.
- Pisatin phytoalexin synthesized by Shikimate and Acetate malonate pathway.
- Phaseollin phytoalexin synthesized by Shikimate, Acetate-malonate and Acetate-mevalonate pathways.

The enzyme **Phenylalanine ammonia lyase** (PAL) is involved in the phytoalexin biosynthesis.

12.2 Role of Phytoalexins in Disease Resistance

The role of phytoalexins in disease resistance varies from plant to plant and even among each plant-pathogen interaction. The following conditions have been reported in various plant-pathogen interactions.

- Phytoalexins are produced in both resistant and susceptible interactions but more in the **resistant** ones.
- Phytoalexins are produced more rapidly in resistant reactions than in the susceptible ones.
- Incompatible pathogens induce more phytoalexins than the compatible pathogens.
- Phytoalexins are degraded by compatible pathogens and not by incompatible ones.
- Phytoalexins are degraded by both pathogens and non-pathogens but rapidly by pathogens.
- Phytoalexins are degraded by the susceptible host itself.
- Non-pathogens are more sensitive to the phytoalexins than the pathogens.
- Pathogens adopt to the phytoalexins quickly compared to non-pathogens.
- Phytoalexins are produced at the appropriate sites at appropriate concentrations and time in the resistant reactions.

The existence of any one of the factors can confer resistance to plants.

12.3 Evidence for Role of Phytoalexins in Defense Mechanism

In 1981, the scientist Keen, described several evidences that support the role of phytoalexins in disease resistance.The evidence included data that is documented:

- Localization and timing of phytoalexin accumulation in infected tissue in relation to pathogen development.
- Strong positive correlation of rapid phytoalexin production with incompatible interactions in gene-for-gene plant pathogen systems.
- Association of rapid phytoalexin accumulation with resistance genes that conditions rapid restriction of pathogen development.
- Use of metabolic inhibitors that enhances susceptibility, block phytoalexin production.
- A positive relationship between pathogen virulence and tolerance of phytoalexins.
- An increase of plant tissue resistance by stimulation of phytoalexin production prior to inoculation.

12.4 Elicitor and Suppressor of Phytoalexins

12.4.1 Elicitor

The term elicitor was first used by Keen et. al. (1972). Phytoalexin production is stimulated in a host by the presence of certain pathogen substances called **elicitors**. Most phytoalexin elicitors are generally high molecular weight substances that are constituents of the fungal cell wall. They are mostly glycoproteins.

12.4.2 Suppressor

The fungi which produce elicitors also produce suppressors. Suppressors, suppress the phytoalexin synthesis in hosts. The suppressors are also mostly glycoproteins. Many phytoalexins have been isolated and characterized from host plants. Some of them are listed in Table 12.1.

Table 12.1

S.No.	Phytoalexin	Host	Pathogen
1.	Rishitin	Potato tuber	*Phytophthora infestans*
2.	Gossypol	Cotton	*Verticillium albo-atrum*
3.	Pisatin	Pea	*Monilinia fructicola*
4.	Phaseolin	Beans	*Sclerotinia fructigena*
5.	Capsidol	Pepper	*Peronospora trifoliorum*
6.	Ipomeamarone	Sweet potato	*Ceratocystis fimbriata*

12.5 Rhizospheric Microbes

Rhizosphere of plants in general contain a large number of microorganisms. In legume crops, the rhizosphere contains a large population of *Rhizobium,* most of which are highly specific to their host other than increasing yield and growth. There are several reports, which show that different species of *Rhizobium* are capable of protecting their hosts against fungal pathogens. The reduction in disease intensity of a particular plant due to the activity of one or more microorganisms in the soil may be attributed to biochemical changes induced in the host. The most important biochemical changes in plants after infection by pathogen are production of phytoalexins.

The role of pisatin tolerance and degradation in the virulence of *Nectria haematococca* on peas was studied by Tegtmeier and Van Etten (1982). A series of crosses were made between isolates of *Nectria haematococca* mating population VI that differed in sensitivity to the pea (*Pisum sativum*) phytoalexin, pisatin. The progeny were tested for sensitivity to pisatin, ability to demethylate pisatin, and virulence on pea. The progeny from some crosses fell into two discrete classes of sensitivity. All progeny in the most tolerant class were able to demethylate pisatin; none in the more pisatin-sensitive class could demethylase this phytoalexin. Tetrad analysis of one of these crosses indicated that at least two loci conferred the tolerant, demethylating phenotype. The progeny from two other crosses did not segregate into distinct classes. In these crosses pisatin sensitivity segregated polygenically, and some of the most sensitive progeny were able to demethylate pisatin. In four crosses, virulence on pea segregated in the progeny. Of the 345 progeny analyzed, all of the moderately or highly virulent progeny were tolerant of pisatin and able to demethylate it. Therefore, either pisatin tolerance and demethylating ability are required for virulence on pea or genes for pisatin tolerance and demethylating ability are required for virulence on pea or genes for pisatin tolerance and demethylating ability are closely linked to genes from virulence. The results of these studies provide, the strongest evidence to date that pisatin tolerance ability to degrade pisatin can determine the virulence of plant pathogen.

Induced resistance and phytoalexin accumulation in biological control of *Fusarium* wilt of Carnation by *Pseudomonas* sp. Strain WCS417r was reported by Van Peer et. al. (1991). The involvement of mechanism other than competition for iron in biological control of *Fusarium oxysporum f. sp. dianthi* of carnation by *Pseudomonas sp.* Strain WCS417r was investigated. Pathogen and antagonist were spatially separated by bacterizing the roots while inoculating the stem with the pathogen. When the roots were bacterized with strain WCS417r one week prior to stem inoculation with *F. o. dianthi*, the number of diseased plants in all experiments with cultivar

'Pallas' was reduced from 50 to 20% and in one experiment with cultivar Lena from 69 to 38%. Strain WCS417r could not be isolated from stem tissue and control of *Fusarium* wilt by strain WCS417r, and therefore, is not due to competition between *F .o. dianthi* and strain WCS417r, but has to be ascribed to induced resistance. Along with induced resistance, there was an increased accumulation of phytoalexins in stems of bacterized and inoculated plants compared with non-bacterized, fungus-infected plants. No accumulation of these compounds was found when the plants were bacterized but were non-infected. The results of these studies concluded that signals provided by strain WCS417r at the root system, induce in the stem sensitization of defense responses against *F .o. dianthi*, such as synthesis and accumulation of phytoalexins.

Occurrence of phytoalexins and phenolic compounds in endomycorrhizal interactions and their potential role in biological control was investigated by Morandi (1996). It has been observed that phytoalexins and associated molecules accumulate in roots after mycorrhizal infection, but less intensively and slowly than in pathogenic interactions. Following mycorrhizal infection, enzymes of phenyl propanoid metabolism have been shown to be activated differentially. Some flavonoids and isoflavonoids have been reported to stimulate *in vitro* germination of mycorrhizal fungi or *in vitro* mycorrhizal infection. A few data available on the occurrence of phenolics in mycorrhizal plant-pathogen interactions support this hypothesis, but much more information is necessary before the role of phytoalexins and phenolics in the biological control of disease in mycorrhizae can be fully understood.

Isolation of phytoalexins from groundnut leaves produced in response to infection with *Cercospora arachidicola* was reported by Sabitha Rani and G.M. Reddy (1999). Phytoalexins were isolated from groundnut leaves, which are induced in response to inoculation with *Cercospora arachidicola*. The pre-inoculative phytoalexin synthesis was more in resistant leaves compared to moderately resistant and susceptible ones. In all the genotypes, synthesis of phytoalexins started within four hours and optimum phytoalexins concentration was obtained after 12 hr. in resistant leaves and 36 hr. in moderately resistant and susceptible leaves of inoculation with the pathogen. These studies clearly indicate that the rate of synthesis and accumulation of phytoalexins plays an important role in making the plant resistant or susceptible to the pathogen.

Induction of resistance to *Penicillium digitatum* in Grapefruit by the yeast biocontrol agent, *Candida oleophila* was studied by Droby et. al. (2002). The yeast, *Candida oleophila*, the base of the commercial product **Aspire**, is recommended for the control of post-harvest decay in citrus and pome fruit. Its modes of action include nutrient competition, site exclusion, and direct mycoparasitism. In the present study, they showed that application of *Candida oleophila* to surface wounds or to intact **'Marsh Seedless'** grapefruit elicited systemic resistance against *Penicillium digitatum*. The induction of pathogen resistance in fruit that was already pronounced 24 hr. after elicitation, was concentration and time dependent and restricted to the peel tissue closely surrounding the yeast application site. The induction of pathogen resistance required viable yeast cells at concentration of 10^8 to 10^9 cells/ml. Application of *Candida oleophila* cell suspensions to grapefruit peel tissue increased ethylene biosynthesis, phenylalanine ammonia lyase activity, and phytoalexin accumulation and increased chitinase and β-1, 3-endoglucanase protein levels. The spore germination and germ tube growth of *Penicillium digitatum* were markedly inhibited in wounds made near the yeast-treated sites. These studies provide evidence that yeast-mediated defense reactions play a supporting role in restricting the spread of fungal infection in wound sites as well as in intact tissue.

Protection of soybean root rot by *Bradyrhizobium japanicum* and *Trichoderma harzianum*, was associated with changes in enzyme activities and phytoalexin production was reported by Chakraborty et. al. (2003). Ten cultivars of soybean were tested for their disease reactions against *Fusarium oxysporum*, the causal agent of root rot disease. The different cultivars exhibited varying degrees of susceptibility with Soymax being the most and JS-2 and UPSM-19 being the least resistant. Seed bacterization with *Bradyrhizobium japanicum* reduced root rot intensity significantly. Application of *Trichoderma harzianum* to soil also reduced root rot intensity. Combined application of *B. japanicum* and *T. harzianum* gave the most significant disease reduction. *B. japanicum* did not exhibit any antagonistic reaction against *F. oxysporum in vitro*, whereas *T. harzianum* inhibited growth of *F. oxysporum*. Phenyl alanine ammonia lyase and peroxidase activities were assayed in both resistant and susceptible cultivars following the different treatments. Activities were significantly higher in the infected roots in comparison to healthy ones. PAL activity was higher in the resistant cultivar but bacterization with *B. japanicum* prior to inoculation with *F. oxysporum* enhanced PAL activity in both the cultivars. Peroxidase activity did not show any increase following pre-inoculation with *B. japanicum*. Glyceollin accumulation, which was significantly higher in the resistant cultivar, also registered a marked increase due to pre-inoculation with *B. japanicum*. *T. harzianum* did not affect enzyme activities or glyceollin accumulation.

Chapter – 13

Mycorrhizae – A Tool in Biological Control

The rhizosphere of plants provide an ecological niche for many of the microorganisms that are present plenty in the soil. Mycorrhiza (pl.mycorrhizae) are structures formed by association of fungi and roots of plants. The term 'mycorrhiza' was introduced by a German botanist, **Albert Bernard Frank** in 1885 which literally means **'fungus root'**. Majority of the plant species form mycorrhizal associations of one kind or the other with certain fungi in soil. These mycorrrhizal fungi perform the function of root hairs. The structure and composition of mycorrhizae are highly variable. Though numerous fungi form inanimate contact with plant roots, only a limited species enter into symbiotic association. Although many of the host cells are invaded by the fungal endophytes there is neither visible tissue damage nor low degree of pathogenecity towards its host.

13.1 Types of Mycorrhizae

Mycorrhizae are classified into three categories namely ectotrophic, endotrophic and ectendotrophic. In ectotrophic mycorrhiza, the fungus forms a mantle or sheath around the rootlets and also enters the root forming an inter cellular net of hyphae called the **'Hartig net'**. The hyphae of Hartig's net do not form haustoria. The predominant types of **mycorrhizae** are endotrophic in nature. In this, the fungus invades the host cells and forms hyphal knots in the cells of root cortex. A portion of the fungus remains in the soil but no external sheath is formed. While mycorrhizae sharing characteristics of both ecto and endotrophic mycorrhizae are called ectendotrophic. As in the case of ectotrophic mycorrhiza, the fungus forms a hyphal mantle and 'Hartig net' and also form haustoria and hyphal coils in the epidermal and cortical cell as do the endotrophic mycorrhiza.

E.g. *Monotrapa indica* (Indian pipe), one of the best studied non-chlorophyllous plant which grows under shades of forest trees, stood example of ectendotrophic mycorrhiza.

13.1.1 Ectotrophic Mycorrhiza

These are commonly occur in temperate regions, members belonging to Basidiomycotina. Mostly found in association with forest trees viz., pines, eucalyptus, oak tree etc. They can be easily detected as the fungus mantle gives the roots a swollen and stumpy look while the uninfected rootlets remain unbranched and short lived. The mycorrhizal fungi are not easy to grow in pure culture and the fruit bodies are especially difficult to induce. Melin et. al. (1953) observed that they require exogenous supply of aminoacids especially thiamin and vitamins. Unlike endotrophic mycorrhiza, they have poor competitive saprophytic ability and grow very slowly in culture.

E.g.

Hymenomycetes : *Entoloma, Cantharellus, Boletus, Amanita, Russula* etc.

Gasteromycetes : *Rhizopogan, Scleroderma*

Ascomycetes : *Tuber, Elaphomyces*

13.1.2 Endotrophic Mycorrhiza

Based on nature of fungi, endotrophic mycorrhizas are classified into groups, namely, (i) those produced by septate fungi belong to Basidiomycotina but rarely Deuteromycotina e.g., *Armillaria mellea* (ii) those produced by aseptate fungi belong to Zygomycotina. The vesicular arbuscular (VA) type of mycorrhiza is the most common among all mycorrhizas. This is the most common type occuring in all ecological situations, especially in tropics. It is well documented that it increased plant growth because of VAM colonization (Bagyaraj and Varma, 1995). The increased plant growth is attributed to enhanced uptake of limited nutrients, hormone production, biological nitrogen fixation, drought resistance and suppression of root pathogens. Formation of two characteristic organs namely **arbuscles**, which are dichotomously branched and **brush like haustoria** and **vesicles**, which are formed inter and intracellularly, form an important feature of VA mycorrhizas. E.g. *Endogone, Rhizophagus.* Unlike the fungi of ectotrophic mycorrhiza, these fungi are capable of utilizing complex carbohydrates like cellulose and lignin in soil. Consequently, the fungi have good competitive saprophytic ability and are successful soil inhabitants.

Biological control of plant pathogens is currently accepted as a key practice in sustainable agriculture because it is based on the management of a natural resource i.e. certain rhizosphere organisms. Biological control can be defined as the directed, accurate management of common components of ecosystems to protect plants against pathogens. Thus, biological control preserves environmental quality by reduction in chemical inputs and is characteristic of sustainable management practices. Several workers have reported that VAM fungi can act as biocontrol agents for alleviating the severity of disease caused by root pathogenic fungi, bacteria and nematodes.

13.2 Vesicular Arbuscular Mycorrhiza (VAM)

These fungi are obligate biotrophs, geographically ubiquitous and occur over a broad ecological range. They are commonly found in association with agricultural crops, shrubs, most tropical tree species and some temperate tree species. It has been observed in 1000 genera of plants representing some 200 families. There are at least 3 lakh receptive hosts in the world flora and 150 species of VAM fungi (Bagyaraj, 1996).

According to **Gerdemann (1975)** it is easier to list the plant families that do not form VAM than to list those that do. Most of the VAM fungi have a broad host range. The fungi associated are non-septate zygomycetous fungi belonging to the genera *Glomus, Gigaspora, Acaulospora, Entrophospora, Sclerocystis* and *Scutellospora.* Many are cosmopolitan in distribution, but some may be strictly tropical like *Acaulospora foveata* and *A. tuberculata.*

The improved plant is attributed to increased nutrient uptake, especially phosphate due to the exploration by the external hyphae of the soil beyond the root hair zone. The phosphate uptake is more significant in soil's deficient in phosphorus. The increased growth of plants inoculated with VAM fungi is not only attributed to improved phosphate uptake but also better availability of other elements like Zn, Cu, Mg, Fe, Mn etc. Allen et. al. (1982) illustrated that VAM affects directly the levels of plant hormones like cytokinins and gibberellin-like substances. Plants colonized by VAM fungi can tolerate a wide range of soil regimes and also improve water relations of many plants.

Mycorrhizal colonization also allows introduced populations of beneficial soil organisms like *Rhizobia, Azotobacter, Azospirillum* and phosphate solubilizing bacteria to maintain higher numbers than in non-mycorrrhizal plants and to exert synergistic effects on plant growth (Bagyaraj, 1989). It is apparent from the investigations of VAM-plant pathogen interactions that VAM can usually deter or reduce the severity of diseases caused by soil-borne pathogens.

13.3 Mechanisms of Suppression of Root Pathogens by VAM Fungi

The principal mechanisms through which suppression of root pathogens occur are morphological, physiological and biological alterations in the host.

13.3.1 Morphological Changes

Colonization of VA mycorrhiza induces remarkable changes in morphology of root system as well as in the meristematic and nuclear activities of root cells (Atkinson et. al. 1994). This might affect rhizosphere interactions and particularly pathogen-infection development.

- In preventing penetration and growth of pathogens like *F. oxysporum* and *Phoma terrestris*, thickening of cell walls through lignification and other polysaccharides were observed in mycorrhizal plants. A strong vascular system observed in mycorrhizal plants will increase the flow of nutrients, impart greater mechanical strength and diminish the effect of vascular pathogens.
- Galls in mycorrhizal plants had fewer giant cells or syncitia, which are essential for the development of nematode larvae when compared to non-mycorrhizal plants, when histopathological studies were conducted on nematode galls caused by *Meloidogyne* spp.

The nematodes in mycorrhizal plants were smaller and took a longer time to develop into adults (Bagyaraj, 1996). Some sort of competition for space occurs when both fungal root pathogens and VAM fungi colonizing the same host cells, usually develop in different root cortical cells, illustrated by Dehne (1982). The *Phytophthora* sp. does not penetrate cortical cells containing arbuscle of VAM was observed by Cordier et. al. (1996). As it suggests that localized competition for infection / colonization site does occur between the pathogen and the VAM fungus.

An increase in branching which further results in relatively larger proportion of higher order roots in the root system is the most frequent consequence of VAM colonization. However, the significance of this finding for plant protection has not yet been sufficiently considered.

13.3.2 Biological Changes

The plants being colonized by mycorrhiza will harbour higher population of microorganisms in the rhizosphere thus making it difficult for the pathogen to compete and gain access to the root (Secilia and Bagyaraj, 1987). Further, they have also isolated more pathogen antagonistic actinomycetes from the rhizosphere of VAM plants than from non-mycorrhizal controls. This is an effect attributed due to the involvement of VAM fungus.

The prophylactic ability of some VAM fungi could be exploited in association with other rhizosphere microorganisms known to be antagonistic to root pathogens that are being used as biological control agents (Linderman, 1994; Azcon Aguilar and Barea, 1996). Preliminary studies suggest that microbial antagonists of fungal pathogens, either fungus or plant growth promoting rhizomicroorganisms (PGPR), do not antagonize VAM fungi. Moreover they can improve the development of the mycosymbiont and facilitate VAM formation. This has been reported especially for *Trichoderma* spp. and *Pseudomonas* spp. Recently, certain bacteria have been shown to stimulate colonization by VAM fungi. These are named as **'mycorrhiza helper bacteria' (MHB)**. Inoculation of neem seedlings with a VAM fungus plus MHB was found to significantly improve growth and nutrition compared to those inoculated with VAM fungus alone. Therefore, the management of these interactions improving plant growth and health, in an integrated approach should form one of the targets for future research.

The microorganisms producing siderophores, which are low molecular weight chelating agents that have high affinity for ferric iron and thus fungistatic to many pathogens were observed in higher numbers in the rhizosphere of mycorrhizal plants. A reduction in *Fusarium* population in the soil surrounding mycorrhizal tomato roots as compared to non-mycorrhizal control was reported by Caron (1989). The number of sporangia and zoospores formed by cultures of *Phytophthora cinnomomi* was reduced by the application of extracts of rhizosphere soil from VAM plants (Meyer and Linderman, 1986). These studies merit further attention in order to elucidate how microbiota-mediated changes may be involved in biological control by VAM association.

13.3.3 Physiological and Biochemical Changes

Research work on VAM fungi revealed that several physiological and biochemical alterations tend to occur due to colonization of mycorrhizas, such as :

(i) improved nutrient status of the host plants,

(ii) higher levels of phenols and aminoacids,

(iii) competition for host photosynthates, and

(iv) activation of plant defense mechanisms.

Vigorous growth of plants results due to increased nutrient uptake and the plant itself may thus be more resistant or tolerant to pathogen attack with onset of VAM symbiosis.

Compared to non-mycorrhizal plants, there has been increased concentration of **ortho-dihydroxy phenols** present in mycorrhizal plants, which was found inhibitory to *S. rolfsii,* a root rot pathogen. Reduction in levels of chlamydospore production in *Thielaviopsis basicola* was reported due to higher levels of aminoacid **'arginine'** in the root extracts of mycorrhizal plants (Sharma et. al. 1992). Tomato roots inoculated with *Glomus fasciculatum* found inhibitory to root nematodes due to increased production of phenylalanine and serine in tomato roots (Bagyaraj, 1996).

Colonization of a plant by VAM fungi alters the host physiology and in turn the root exudation pattern. Due to the decreased root exudation in mycorrhizal plants, which in turn is responsible for increased membrane phospholipid content possibly helps in reducing the infection by root pathogens. It has been proposed that the growth of both the VAM fungi and root pathogens depends on host photosynthates and that they compete for carbon compounds reaching the ground. When VAM fungi have primary access to photosynthates, the higher carbon demand may inhibit pathogen growth. However, there is little or no evidence that competition is a generalized mechanism for pathogen biocontrol activity of VAM symbiosis.

Activation of specific plant defense mechanisms as a response to VAM colonization is an obvious basis for the protective capacity of VAM fungi. The compounds involved in plant defense in relation to VAM formation are phytoalexins, chitinases, β-1, 3-glucanases, peroxidases, pathogenesis-related proteins, hydroxy proline-rich glycoproteins (HRGP) and phenolics. Electrophoretic analysis of soluble extracts from VAM roots has demonstrated that the host plant produces a number of new proteins, namely, **'endomycorrhizins'** in response to VAM colonization. The low molecular, toxic compounds, namely, 'phytoalexins' usually accumulate with pathogen attack and released at the site of infection. These are not detected during the first stages of VAM formation but can be found at later stages of the symbiosis (Morandi and Gianinazzi et. al. 1984). This was true for phytoalexin **'glyceollin'** of soybean and **'medicarpin'** of *Medicago trancatula*.

There is increased activity and amount of phenyl propanoid-related enzymes such as **phenyl alanine ammonium-lyase (PAL)**, the first enzyme of the phenyl propanoid pathway and chalcon isomerase, the second enzyme specific for flavanoid/isoflavanoid biosynthesis, during early colonization of plant roots by *Glomus intra- radices* but sharply decreased to levels at or below those in uninoculated controls. These results suggest that VAM fungi initiate a host defense response, which is subsequently suppressed.

From the results, it is evident that VAM colonization will activate plant defense mechanisms to a very little extent. These compounds could sensitize the roots to pathogens and enhance mechanisms of defense to subsequent pathogen infection (Azcon Aguilar and Barea, 1996). Current research using molecular biology techniques, immunological and histochemical analyses will provide more information about these mechanisms.

13.4 VAM in the Management of Root Pathogens

There are several reports indicating the decreased disease severity or reduction due to VAM colonization. Integrated application of VAM and castor cake in the management of root-knot nematode on egg plant (aubergine) have shown that seedlings colonized with mycorrhiza were least infected by *M. incognita* when transplanted in soil which was amended with castor cake (Rao et. al. 1998). Significant increase in colonization of *Glomus fasciculatum* on roots of aubergine and chlamydospore densities of mycorrhiza indicated favourable effects of castor cake amendment on the growth of *G. fasciculatum*.

Similarly, integration of the VAM fungus *Glomus mosseae* and oil cakes of castor, pongamia and margosa showed that integration of VAM with margosa cake reduced *M. incognita* population in soil and reduced root galling, egg mass production and fecundity and produced vigorous tomato seedlings with increased root colonization by *G. mosseae* (Reddy et. al. 1998).

Another study with staggered inoculation of VAM and root pathogen brought out that application of VAM two weeks earlier than pathogen inoculation was better than simultaneous application of VAM and the pathogen in reducing the severity of disease. Application of pathogen two weeks earlier than VAM was not effective in controlling the disease (Suresh and Bagyaraj, 1984). The interaction of VAM of Arabica coffee and chemical control of nematodes was studied by Vast (1997). It was found that when VAM fungi was inoculated well in advance, it resulted in an enhanced tolerance of Arabica coffee cultivars which are susceptible to endoparasitic nematodes viz., *Pratylenchus* and *Meloidogyne* that are widespread in Arabica plantations.

Consistent reduction of disease symptoms with VAM fungi has been described for fungal pathogens such as *Phytophthora parasitica,*

Gaeumannomyces graminis var. *tritici, Fusarium oxysporum, Rhizoctonia solani, Sclerotium rolfsii, Pythium ultimum* and bacteria such as *Pseudomonas syringae* and *P. solanacearum* and nematodes such as *Meloidogyne* spp. *Tylenchulus semipenetrans, Pratylenchus brachyurus* and *Radopholus similis*.

From the above studies, the main conclusion that can be drawn are:

- VAM associations can reduce damage caused by soil-borne plant pathogens.
- The abilities of VAM symbiosis to enhance resistance or tolerance in roots are not equal for the different VAM fungi so far tested.
- Protection is not effective for all pathogens.
- Protection is modulated by soil and other environmental conditions.

Chapter – 14

Plant Growth Promoting Rhizobacteria

Fluorescent pseudomonads are the major bacterial group found surviving in rhizosphere of all field crops. They are also known as '**Plant Growth Promoting Rhizobacteria' (PGPR)** as they promote plant growth by secreting auxins, gibberellins and cytokinins (Lifshitz et. al. 1987). They survive in a wide variety of habitats such as soil, rhizoplane, phylloplane, pistils, nectarines and fruits of plants. *Pseudomonas fluorescens, P. putida* and *P. cepacia (Burkholderia cepacia)* are the important fluorescent pseudomonad species which are known for their high biocontrol efficacy. The other group of bacteria which come under the category of PGPR is spore forming bacteria, *Bacillus.*

Plant growth promoting rhizobacteria has unquestionable potential for managing plant diseases and increases the crop productivity. Introduction of fluorescent pseudomonads as biocontrol agents offers a promising alternative to manage soil-borne plant pathogens. Roots can be attacked by soil-borne plant pathogens, which cause diseases such as root rot and vascular wilt. Although alternative techniques such as crop rotation, steam treatment and solarization of soils are available to reduce damage to crops by phytopathogenic organisms, farmers generally use chemical agents for economical reasons. However, non-target effects of these chemicals (e.g. methyl bromide) raise concern for the environment and public health. Introduction of bacterial and fungal biocontrol agents offers a promising alternative to manage soil-borne diseases. During the past 20 years, many biocontrol organisms have been isolated from soils naturally suppressive to certain soil-borne plant pathogens. To date many bacterial genera are being used and tested in bacteriazation, including *Acinetobacter, Agrobacterium Alcaligenes, Arthrobacter, Azospirillum, Azotobacter, Bacillus, Bradyrhizobium, Frankia, Pantoea, Serratia, Pseudomonas, Rhizobium, Stenotrophomonas, Streptomyces* and *Thiobacillus* (Weller, 1988). Many biocontrol agents have been identified as root-colonizing bacteria of the genus *Pseudomonas.* In general, biocontrol psudomonads are not specific for one pathogen plant species only, but have a wide host range and suppress several pathogens.

14.1 Mode of Action of PGPR

In the past years, many research groups have focused on the modes of action of biocontrol and several mechanisms have been described that are responsible for the disease-suppressive capacity of these pseudomonads. They suppress the pathogen directly by antagonism through the production of various secondary metabolites or indirectly by inducing plant mediated defense reactions. (Fig. 14.1).

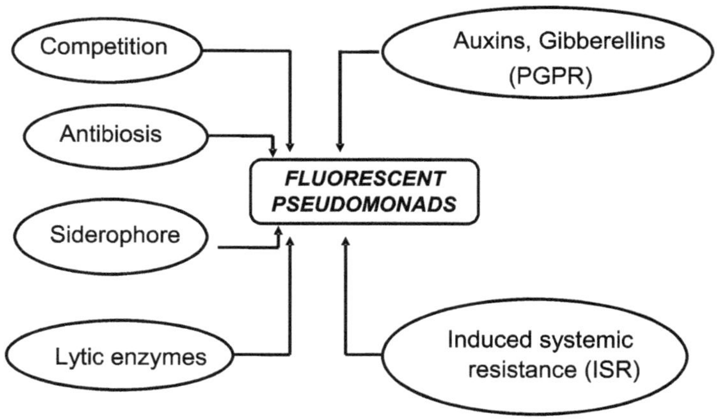

Fig. 14.1 Mechanisms of action exhibited by fluorescent pseudomonads.

14.1.1 Root Colonization

Efficient root-colonization is a prerequisite for a strain to be a good biocontrol agent. The biocontrol bacteria must be able to establish itself at population densities sufficient to cause a beneficial effect. It has to survive in the rhizosphere, proliferate, colonize the entire root system efficiently and be able to compete with indigenous microorganisms. It has been shown that a mutant of *P. chlororaphis* PCL 1391, impaired in colonization traits only was no longer able to protect tomato from foot and root rot caused by *Fusarium oxysporum* f. sp. *radicis – lycopersici*. Once the root is successfully colonized, plant beneficial pseudomonads may produce several extracellular factors, which are suggested to play an important role in disease suppression Siderophores, extracellular enzymes, compounds that induce plant defense mechanisms. None of these mechanisms are mutually exclusive and frequently several modes of action are exhibited by a single biocontrol agent. Fluorescent pseudomonads are efficient root colonizers because they occur in the natural habitat of rhizosphere and thus when they are re-introduced through seed or seed-piece inoculation, they colonize root surface profusely. (Fig. 14.2).

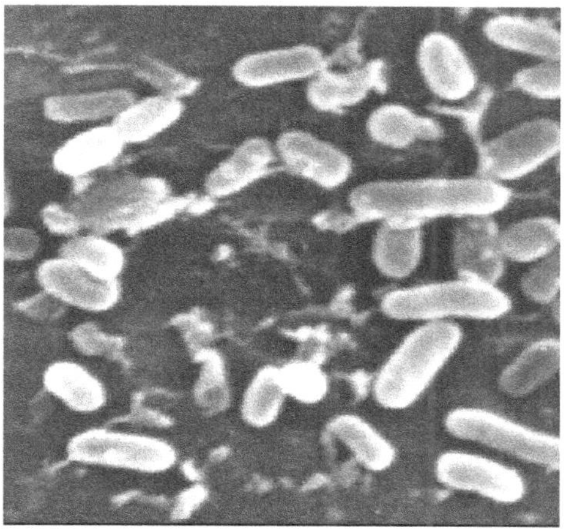

Fig. 14.2 Rhizosphere colonization by pseudomonas fluorescens.

14.1.2 Siderophores

Siderophores are extracellular, low molecular weight substances which selectively complex iron with high affinity. Competition between bacteria and fungal plant pathogens for space and nutrients is believed to be an important biocontrol mechanism. In the past years, many studies have been focusing on biological control by rhizosphere inhabiting fluorescent pseudomonads through competition of iron. Under iron-limiting conditions, bacteria produce a range of iron-chelating compounds or siderophores, which have a very high affinity for ferric iron. These bacterial iron chelators are thought to sequester the limited iron supply available in the rhizosphere making it unavailable to pathogenic fungi and thereby restricting their growth. Several bacterial siderophores are associated with biocontrol of soil-borne pathogens. For example, **"Pyoverdin"** and **'Pyochelin'** reportedly play a role in the suppression of *Pythium* induced damping-off of tomato by *Pseudomonas aeruginosa* 7NSK2. Fluorescent pseudomonads produce siderophores such as **'pseudobactin'** and **'pyoverdine'**, which chelate the iron available in the soil and make it unavailable to pathogen, thus the pathogen dies for want of iron.

14.1.3 Extracellular Enzymes

Production of lytic enzymes by biocontrol agents is another factor which is associated with parasitization of fungal hyphae. Lytic enzymes such as chitinases and β-1, 3-glucanases are produced by fluorescent pseudomonads which degrade chitin and glucan fragments of fungal cell wall respectively. Ultra structural and morphological changes of mycelium of *Verticillium dahliae* and inhibition of mycelial growth were observed due to lytic enzyme production by *P. fluorescens*. Culture filtrate of *P. fluorescens* grown in chitin-amended medium, which is substrate for the production of chitinase, significantly inhibited the mycelial growth of *Colletotrichum falcatum* causing red rot of sugarcane *in vitro*. Positive correlation was observed in chitinase production by isolates of *P. fluorescens* and suppression of sheath blight incidence in rice caused by *R. solani*.

Another example is *Serratia marcescens*, a biocontrol agent found to possess chitinolytic activity, able to release 'N-acetyl D-glucosamine' from cell walls of *Sclerotium rolfsii*. When *S. rolfsii* was sprayed with partially purified chitinase from *S. marcescens,* rapid and extensive bursting of the hyphal tips was observed. This chitinase preparation was effective in reducing disease incidence caused by *S. rolfsii* in bean and *R. solani* in cotton under greenhouse conditions.

14.1.4 Induced Resistance

All plants have natural defense mechanisms against pathogen attacks. A virulent pathogen avoids triggering these mechanisms, suppresses resistance reactions or evades the effects of active defenses. Disease can be reduced if these mechanisms are triggered by leaf-necrosis pathogens or rhizobacteria prior to infection and this phenomenon is commonly known as **systemic acquired resistance (SAR)** and **induced systemic resistance, (ISR)** respectively. Various non-pathogenic rhizobacteria have been shown to induce systemic resistance in plants and thereby providing protection against a broad spectrum of phytopathogenic fungi, bacteria and viruses. Several bacterial determinants are claimed to produce systemic resistance and include siderophores, salicyclic acid and the O-antigenic side chain of the bacterial outer membrane protein lipopolysaccharide (LPs).

Colonization of tobacco roots by *Pseudomonas fluorescens* CHAO reduces leaf necrosis caused by tobacco necrosis virus (TNV) and induces physiological changes in the plant (e.g. increase in salicylic acid and PR proteins in leaves). A pyoverdine-negative mutant of CHAO could only partially induce resistance against TNV (Mauhofer et. al. 1994). Salicylic acid production is essential for induction of resistance to *Botrytis cinerea* by *P. aeruginosa* 7NKS2 in bean even in very small amounts. In the systemic protection of radish against Fusarium wilt by *P. fluorescens* WCS 374 or WCS 417, the bacterial LPS appeared to be the trait responsible for resistance induction.

14.1.5 Antibiosis

Fluorescent Pseudomonads suppress the pathogens by antibiosis through the production of various antibiotic substances such as 2,4-diacetyl phloroglucinol, phenazine -1-carboxylic acid, oomycin, oxychlororaphine, pyoluteorin, pyrrolnitrin and pyocynine. It is now well established that the production of antimicrobial compounds is one of the most important mechanisms in biocontrol. The polyketide antimicrobial metabolite, DAPG has emerged as one of the most important antimicrobial compound produced by biocontrol strains of *Pseudomonas fluorescens*. Application of purified 2,4-diacetyl phloroglucinol inhibited the infection of *Pythium* in sugar beet (Vincent et. al. 1991). *P. fluorescens* strain 5-2/4 produced the maximum amount of 2,4-diacetyl phloroglucinol and protected tomato seedling from damping off incidence caused by *P. ultimum* when compared to wild strain. (Table 14.1).

Table 14.1 Antibiotics produced by PGPR.

PGPR	Antibiotics
Pseudomonas fluorescens (CHAO)	Pyrrolnitrin
	Pyoluteorin
	Chlororaphine
	Oxychlororaphine
	2,4-diacetylphloroglucinol
P. fluorescens MMB-16	Aerugine
Pseudomonas sp	Antifugal antibiotics
	Phenazines
	Pyrrolnitrin
	Pyoluteorin
	2,4-diacetylphloroglucinol
	Rhamnolipids
	Oomycin A
	Capaciamide A
	Ecomycins
	DDR
	Viscosinamide
	Butyrolactones
	N-butyl benzene sulphonamide
	Antibacterial antibiotics
	Pseudomonic acid
	Azomycin
	Antitumour antibiotics
	FR901463

Table 14.1 Contd...

PGPR	Antibiotics
	Cepafungins
	Antiviral antibiotic
	Karalicin
Bacillus sp.	Kanasamine
	Zweittermycin A
	Luturim A (Cyclopeptide)
	Surfactin

In wheat, suppression of take all of wheat caused by *Gaeumannomyces graminis* var. *tritici* was mainly due to the production of phenazine-1-carboxylic acid (Thomashow and Weller, 1988). Insertion of PCA gene, which encodes for the synthesis of phenazine -1-carboxylic acid (PCA) in wild type *P. fluorescens,* significantly improved its ability to reduce damping-off disease of pea seedlings caused by *P. ultimum* and level of biosynthesis was correlated with biocontrol efficacy and the persistence of bacterial colonies in soil ecosystem.

Suppression of black root rot caused by *Thielaviopsis basicola* and take-all of wheat by *P. fluorescens* strain CHAO was mainly due to the production of 2,4 diacetyl phloroglucinol (Keel et. al. 1990). This antibiotic was also inhibitory to *R. solani* and *P. ultimum in vitro*.

Howie and Suslow (1991) reported that reduction in seedling inoculum of *P. ultimum* by *P. fluorescens* HV 37R2 in cotton was mainly attributed to the production of antibiotic **'Oomycin A'** which was found to reduce the secondary inoculum of *P. ultimum.*

14.2 Importance of Antibiotics in Disease Control

Fungicides of microbial origin are synthesized biologically that are specifically effective on the target organisms and inherently biodegradable. Microbial products are so complex in their chemical structures that one compound may have two or more totally different chemical moieties which can interact with different receptors. The complexity in their structures may make natural products chemically and biologically diverse. Since the first success in the use of streptomycin for the control of pear fire blight caused by *Erwinia amylovora* in the United States, many attempts have been made to find microbial products for plant disease control. Eventually, an epoch-making substance, named '**Blasticidin**' active against rice blast disease, was discovered in Japan. Since a variety of antibiotics like kasugamycin, polyoxins, validamycin A and mildiomycin have been commercially used for control of rice blast, rice sheath blight and powdery mildew of rose, respectively. Strong *in vitro* antifungal activities have recently been discovered from the microbial metabolites. Some other antibiotics such as gopalamicin, tubercidin, a manumycin-type antibiotic, oligomycin A, streptimidone, daunomycin and phenyl acetic acid were also found to have a potent antifungal or antioomycete activity for control of economically important plant diseases.

Some pseudomonads have been recognized as antagonists of plant fungal pathogens and antibiotic producers. This is probably due to the abundance of this diverse group of bacteria and their obvious importance in the soils. A single strain of *Pseudomonas* can produce several different antibiotics. In particular, the pseudomonas product '**Pyrrolnitrin**' serves as a lead compound for new fungicides fenpiclonil and fludioxonil. Since PGPR is a potential candidate in disease management through multiple modes of action, it becomes highly imperative to know about the role of antibiotics in the management of plant pathogens.

14.3 Mode of Action of Antibiotics

There are different modes of action reported due to the application of antibiotics in the management of plant pathogens.

- Vacuolization of the nuclear envelop
- Increase in cell wall thickness
- Inhibition of hyphal elongation
- Lysis of inner mitochondrial membrane
- Golgi vesicle trafficking
- Localized alteration of plasma membrane.

14.4 Structural and Physiological Changes

It is evident from several studies that fluorescent pseudomonads induce systemic resistance against many pathogens and control diseases caused by them and strains which induce resistance have been reported to be endophytic in nature (Benhamou et. al. 1996b).

- Penetration and colonization of some host epidermal cells and the outer cortex of tomato plants.
- Frequent occlusion of epidermal cells with a densely stained amorphous material.
- Formation of polymorphic wall thickenings along the host epidermal cells.
- At the onset of bacterial antagonists mediated induced resistance in several plants, marked host metabolic changes culminating in a number of structural changes viz., accumulation of callose and lignin occurred.

- Occurrence of biochemical responses viz., synthesis of chitinases.
- A massive accumulation of phytoalexins could be detected in bacterized carnation roots after pathogen challenge (Van Peer et. al. 1991).
- Accumulation of phenolic compounds in the lumen of host cells due to *P. fluorescens* treatment followed by challenge inoculation with *Fusarium oxysporum radicis* f. sp. *lycopersici* in tomato (M'Piga et. al. 1997). (Fig. 14.3).
- An increase of peroxidase activity as well as an increase in the level of mRNA encoding for PAL and chalcone synthase could be seen in bean roots colonized by bacterial antagonists (Zdor and Andersen, 1997).
- Salicylic acid is a systemic signal molecule inducing disease resistance and fluorescent pseudomonads are known to produce salicylic acid (Dowling and O'Gara 1994).

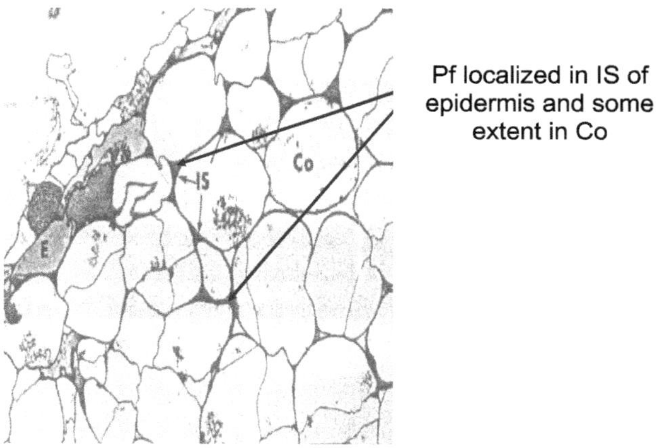

Fig. 14.3 Endophytic nature of P. fluorescens.

14.4.1 Quorum Sensing

'Quorum sensing' is one of the core areas of research on plant microbe interactions. It has been recognized in recent years that bacteria can behave not only as individual cells, but also under appropriate conditions and when their numbers reach a critical level, they can modify their behaviour to act as a multicellular group. This cell-to-cell communication is aided by chemical signals, which is released into the surrounding environment. These chemical signaling molecules are detected and quantified by each cell in order to act as a quorum. The major signals that these bacteria use in quorum sensing is a family of structurally related compounds called **acyl homoserine lactones (AHL).**

14.5 Selection of Potential Strains of Antagonistic Bacteria

Fluorescent Psudomonads are ubiquitous in rhizosphere of crop plants, but only a few of them are effective antagonists. Out of 27 fluorescent Pseudomonad strains tested for their efficacy, only four strains were found inhibitory to chickpea wilt pathogen, *Fusarium oxysporum* f. sp. *ciceris in vitro* (Vidhyasekharan and Muthamilan, 1995). A wide variation among 14 strains of *P. fluorescens* was observed in their ability to inhibit sheath blight of rice caused by *Rhizoctonia solani.* The consistency of biological control against *Colletotrichum orbiculare* in cucumber varied with the strain used (Liu et. al. 1995). These studies indicate that selection of proper strain is very much important in developing biological control strategy.

14.6 *In vitro* Screening Methods for Antagonistic Bacteria

The bacteria isolated from rhizosphere and phyllosphere are to be screened for their efficacy to inhibit the pathogens under *in vitro.* There are commonly three methods viz., dual culture technique, greenhouse inhibition of plant pathogens and test for plant growth promoting ability are to be used for screening.

14.6.1 Dual Culture Technique

The *in vitro* antagonism of isolated bacterial strains toward plant pathogens can be effected by streaking the bacteria in a 4 cm line on each test medium. About 5 mm diameter of mycelial discs or a loopful of bacterial pathogen is to be transferred to the most distal point from the bacteria on the plate. The plates are to be incubated at 28^0C for 6-9 days. After 8 days, the zone of inhibition (in mm) can be calculated by taking the difference of growth in control plate (without bacteria) and the growth of bacteria in treated plate (with bacteria). Three replications are to be maintained for each treatment.

14.6.2 Greenhouse Inhibition of Plant Pathogens

To conform the inhibiting ability of the promising bacterial isolates, this test is usually conducted after initial screening by dual culture under glass house conditions. The methods generally used are (i) Root Dip Method and (ii) Soil Treatment.

14.6.2.1 Root Dip Method

Shake cultures of test bacteria are prepared on Tryptic Soy Agar (TSA) broth for about 24-48 hrs. and estimate colony forming units / ml by serial dilution. The inoculum of the target pathogen for inoculation is prepared. In the case of soil-borne pathogens viz., *Fusarium oxysporum,* a wilt pathogen, it can be grown on wheat bran : vermiculite (1:1) mixture of 15 days and mix with soil at the rate of 10-15 g/kg of soil. Whereas for aerial pathogens, suspension containing spores can be sprayed or swabbed onto plants after 15 days of growth. Pick out the young seedlings of 2 weeks old intact with root system and immerse in the test bacterial solution having a minimum population of 9×10^8 cfu / ml. After immersion in the bacterial suspension, plant the seedlings in pathogen infested soil. Simultaneously, the antagonistic will be allowed to establish in soil before inoculating with pathogen for about 2 weeks. Observe for disease incidence after 15 days and the ability of antagonist can be evaluated by comparing with fungicide treatment.

14.6.2.2 Soil Treatment

Inoculum of test bacteria can be cultured in TSA broth by allowing constant agitation for about 24-48 hrs. and cell density can be estimated by serial dilution. Talc formulations are generally used for soil treatment, at the rate of 2 g / kg of soil or 2.5 kg / ha. Antagonistic ability of bacteria can be estimated by comparing with fungicide treated plots.

14.6.3 Test for Plant Growth Promoting Ability

After initial evaluation of antagonistic ability of promising bacteria by dual culture and based on inhibition zone, they can be further screened for their plant growth promoting ability by comparing the shoot weight, pod weight, root weight, biomass production etc., of treated plants with that of untreated or control plants.

The test plants are to be inoculated with the promising antagonistic bacteria either by seed or soil treatment. Simultaneously, grow the test plants without treatment with antagonistic. Take out observations of root weight, shoot weight etc., after allowing the plants to grow for a period of 15 days. Re-inoculate the treated plants with the test antagonist and observations will be made after 30 days and again after plant maturity for the said parameters. Compare the observations with that of untreated plants and based on observations, the antagonistic will be selected which promote plant growth.

14.7 Inoculum Dose

The antagonist inoculum dose determines the efficacy of antagonist in control of diseases (Johnson, 1994). For significant suppression of *Fusarium* wilt of radish, a threshold population density of fluorescent Pseudomonad strains of approximately 10^5 cfu per g of root is required. When rhizosphere population densities of the strains dropped below this threshold level, efficacy of these strains to suppress the *Fusarium* wilt was almost lost (Raajimakers et al. 1995). The concentration of *Pseudomonas putida* cells in the seed treatment suspension significantly affected their biocontrol efficiency in controlling *Pythium ultimum* infection in pea and soybean (Whipps and Lumsden, 1991). An inoculum dosage of 10^7 to 10^8 cfu per ml of cell suspension was needed for effective control of rice sheath blight pathogen by *Pseudomonas fluorescens* strain Pfl (Vidhya Sekharan and Muthamilan, 1998).

14.8 Method of Application

The antagonistic bacteria are established well in rhizosphere when they are introduced at germinating seedling stage. Hence, fluorescent Pseudomonads should be applied as seed treatment. There are several reports indicating when these bacteria are applied as seed treatment, they are well established in rhizosphere of several crops. Soil application of these bacteria also introduces the bacteria into rhizosphere but less effectively than seed treatment. When *Pseudomonas fluorescens* strain PF1 was applied through seed, the rhizosphere population in was 4.6×10^6 cfu /g, while it was only 0.3×10^6 cfu /g when it was applied through soil (Vidhya Sekharan and Muthamilan, 1998). Seed treatment with the antagonistic bacteria has been reported to be effective in the control of various diseases.

14.8.1 Seed Bio-priming

Bio-priming of treated seeds increased the bacterial antagonists population several fold in seed (Callan et. al. 1990). Bio-priming of chickpea seed after treatment with *P. fluorescens* strains Pfl increased the rhizosphere population of the bacteria by about 100 percent over that with seed treatment only (Vidhya Sekharan and Muthamilan, 1995). Root treatment with the Pseudomonads at the time of transplanting effectively controls rice diseases.

14.9 Time of Application

Seed treatment gives effective protection against pathogens by inducing defense mechanisms much before the pathogen attack. However, the seed treatment offers only transient protection by inducing defense genes. Efficacy of seed treatment is lost within 14 days in bean (Zdor and Anderson, 1992) and 45 days in rice (Vidhya Sekharan et. al. 1997a). Hence, seed treatment may be effective in control of only early infection by pathogens. Since, these bacterial antagonists control diseases mostly by inducing disease resistance, the antagonists must be applied before the pathogen inoculation. They cannot be applied once outbreak of disease is noticed. In a detailed study, Kamala Nayar and Vidhyasekharan (1998) have shown that when *P. fluorescens* was applied even two days after inoculation with *Xanthomonas oryzae* pv. *oryzae* was found ineffective in control of bacterial leaf blight in rice while when the bacterial antagonist applied 4 to 5 days prior to inoculation with the pathogen could effectively controlled the disease. The bacterial antagonist should be sprayed at least 3 to 4 days before inoculation with the rice blast pathogen, to control the disease effectively. These results suggest that time of application of the antagonists is important in management of diseases.

14.10 Development of Formulation

The major problem associated in developing a biocontrol product of bacteria is the achievement of adequate growth. When compared to antagonistic fungi, bacterial agent is more suitable for fermentation-based mass production technology. The problems involved with formulation technology are as important and complex as those associated with growth of biocontrol bacteria.

14.10.1 Ideal Characteristics of a Formulated Product

The following characteristics are ideal for preparing a bioformulation

- Easy to prepare and apply
- Raw materials used should be inexpensive
- Availability of substrates in abundance
- Stability during transportation and storage
- Good shelf-life
- Sustained efficacy
- Availability of product at acceptable cost.

During the development of product, especially in large scale production system, drying and milling can adversely affect the quality of the product.

Large scale slow drying often lead to inactivation of the effective molecule as well as in increase in contaminating microbes. These factors frequently influence quality control and product effectiveness and these should be constantly monitored.

Different methods include seed treatment, root dip treatment, sett treatment in sugarcane, sucker treatment in banana, soil application and foliar application are available for control of diseases. However, combination of different methods of application could be more effective in disease management than a single method of application. Kloepper and Schorth (1981) carried out first the mass multiplication of rhizobacteria in a suitable medium and developed a powder formulation. The suitability of different carriers for the development of stable formulations in talc-based and peat-based formulations was assessed by Vidhya Sekharan and Muthamilan (1995). The other materials tested are vermiculite, lignite and kaolinite as carriers. *P. fluorescens* strains Pf1 and Pf2 survived for more than 180 days in talc-based formulations and 180 days old formulations were as effective as freshly prepared formulations in the control of pigeonpea wilt. Capsule formulations of *P .fluorescens* have been developed and they were found to be effective in the control of banana wilt (Vidhya Sekharan et. al. 1998). Talc-based formulations were found better against various pathogens in several crops than aqueous preparations. Suslow (1980) developed a dried powder formulation of PGPR in which methyl cellulose was used as an adhesive to enhance the efficacy of the formulation.

14.10.2 Improvement of Efficacy of Formulation by Chitin Amendment

To enhance the survival and efficacy of PGPR, several molecules have been added to the medium while developing a formulation. Among them, chitin and chitosan are reported to reduce disease caused by soil-borne pathogens. Chitin would serve as a carbon source or substrate for chitinolytic bacteria. When bacteria were grown in chitin-amended medium, there was increased production of chitinase enzyme observed (Gooday, 1998). In addition, chitin oligomers which are released during degradation of chitin substrate by chitinolytic bacteria are also found to elicit plant defense reactions (Benhamou and Theriaut, 1992). Incorporation of chitin in King's medium B (KM B) supported the multiplication of *Pseudomonas fluorescens* when compared to the medium without incorporation of chitin and enhanced chitinase activity was observed in chitin-amended medium (Viswanathan and Samiyappan, 2001). Recently, Rajda Commare et. al. (2002) demonstrated that chitin – amended PGPR formulation significantly controlled the sheath blight disease and leaf folder in rice besides enhancing the grain yield and attracting more number of natural enemies of leaf folder insect in rice field. Biocontrol efficacy of rhizobacteria (as mixture of P7-1 + FP-7) was enhanced due to induction of defense – related enzymes viz., chitinase, β-1,3 glucanase, peroxidase, PAL, polyphenol oxidase etc., in chitin-amended treatments. Expression of defense related genes in chitin amended treatments contributed more towards durable ISR against fungal pathogen and insect pest simultaneously in rice plants under field conditions.

Chitosan, a non-toxic polymer obtained from the chitin of crustacean shell, is not only the inhibitor of fungal growth but also activates genes encoding defense related proteins in plants (Hadwiger et al. 1986; Lofontaine and Benhamou, 1996). When tomato plants treated with chitosan, enhanced protection against root rot caused by *Fusarium oxysporum* f. sp. *radicis- lycopersici* was reported. In groundnut, treatment of leaves with chitosan before challenge inoculation with *Puccinia* pathogen reduced the rust disease (Sathiyabama and Balasubramanian, 1998).

When zinc sulphate is added in a growth medium, it acts as a growth factor and increases production of antibiotics viz; phenazine-1-carboxylic acid produced by *P. fluorescens* strain 2-79, an important biocontrol agent of take-all fungus pathogen, *Gaeumannomyces graminis* var. *tritici* (Slininger and Jackson,1992).

14.11 PGPR Mixture

The bacterial strains are mostly applied singly to combat a plant pathogen in the biocontrol research so far with rhizobacteria. In many studies, it has been demonstrated the potential benefits in the application of single biocontrol bacterial strain. However, it may also partially account for the reports of inconsistent performance, because a single bioagent is not likely to be active in all kinds of soil environments and agricultural ecosystems (Raupach and Kloepper, 1998).

Associated Problems

- Inadequate colonization by biocontrol agent
- Limited tolerance to changes in environmental conditions
- Fluctuations in production of antifungal metabolites

Several approaches have been used to overcome these problems since biological control results from mixtures of antagonists rather than from high populations of a single antagonist in most of the natural ecosystems. It includes, combined application of two or more biocontrol bacterial strains to enhance the level and consistency in disease control.

In India, multiple strain mixture of bacterial agents have been employed with significant success against a spectrum of plant diseases caused by fungi, virus and nematode pathogens. Nanda Kumar et. al. (2001 a) demonstrated that a PGPR mixture containing *P. fluorescens* strains, Pfl and FP7 showed enhanced protection against sheath blight disease with simultaneous increase in grain yield when compared to individual strains or chemical fungicides or chemical treatment. Similarly, Rajda Commare et. al. (2002) reported the enhanced efficacy of PGPR mixture against the leaf folder in rice besides simultaneous control of *R. solani* pathogen when compared to single strains. In tomato, Kandan et. al. (2002) observed a significant reduction of tomato spotted wilt virus disease with plants treated with PGPR mixture compared to single strains. The PGPR mixture comprising Pfl and FP7 strains have been reported to be the best inducers plant chitinase and peroxidase isoforms which are the most important components of induced systemic resistance.

14.12 Endophytic PGPR

Endophytic PGPR are defined as bacteria, which reside within the living plant tissues without doing substantative harm or gaining benefit other than residency (Kado, 1992). In addition to rhizosphere and rhizoplane colonization, certain rhizobacteria strains are reported to be endophytes localized in the intercellular spaces of root epidermal cells and vascular tissues.

The following diseases are reported to be controlled by PGPR under field conditions.

Table 14.2 Suppression of Major Diseases by Rhizobacteria under Field Conditions

Crop	**Pathogen**
Vegetables	
(i) Tomato	
Wilt	*Fusarium oxysporum f.sp. lycopersici*
(ii) Bean	
Halo blight	*Pseudomonas savastanoi pv. phaseolicola*
Root rot	*Sclerotium rolfsii*
(iii) Pea	
Wilt	*Fusarium oxysporum f.sp. pisi*
Root rot	*Aphanomyces euteiches*
Damping off	*Pythium ultimum*

Table 14.2 Contd...

Crop	**Pathogen**
(iv) Cluster beans	
(i) Bacterial blight	*Xanthomonas axonopodis pv. cyamopsidis*
(v) Cucumber	
Wilt	*Fuasrium oxysporum f.sp. cucumerinum*
Damping off	*Pythium aphanidermatum*
Anthracnose	*Colletotrichum orbiculare*
Angualr leaf spot	*Pseudomonas syringae pv. lachrymans*
(vi) Radish	
Wilt	*Fusarium oxysporum f.sp. raphani*
(vii) Potato	
Soft rot	*Erwinia caratovora*
Wilt	*Ralstonia solanacearum*
Fruit Crops	
(i) Apple	
Grey mold	*Botrytis cinerea*
(ii) Pear	
Fire blight	*Erwinia amylovora*
(iii) Banana	
Wilt	*Fusarium oxysporum f.sp. cubense*
Cereal Crops	
(i) Rice	
Blast	*Pyricularia oryzae*
Sheath blight	*Rhizoctonia solani*
Bacterial blight	*Xanthomonas oryzae pv. oryzae*
Sheath rot	*Sarocladium oryzae*
Tungro	*Rice Tungro Virus*
(ii) Wheat	
Take-all	*Gaeumannomyces graminis var. tritici*
Leaf spot	*Septoria tritici*
Rust	*Puccinia recondita*
(iii) Maize	
Damping off	*Pythium spp.*
Oil seed crops	
(i) Groundnut	
Late leaf spot	*Cercosporidium personata*
Root rot	*Sclerotium rolfsii*

Chapter – 15

Biological Suppression of Weeds with Pathogens

Biological weed control is an approach utilizing living organisms to control or reduce the population of an undesirable weed species. Biological control agents can be specific to the weed and cause neither residue problems nor accumulation of toxic products in soil or underground water. Theoretically any type of plant pathogen viz., virus, bacterium or fungus can be considered as a biocontrol agent for weeds.

There are two main situations in which plant pathogens can be used as weed control agents. The two distinct tactics apparently emerging are:

(i) Classical or traditional or inoculative approach.

(ii) Bioherbicidal strategy.

15.1 Classical or Inoculative Approach

The classical or inoculative approach involves the importation and release of one or more natural enemies that attack the target weeds in its native range. It is released into areas where the weed is introduced and is troublesome and where its natural enemies are absent. The classical strategy is an ecological approach which relies on the ability of an organism to multiply and spread following small scale release.

Objective : The objective of classical biological control is generally not eradication of the weed species but regulation of self-perpetuating weed population at acceptable low levels.

Rust fungi have been studied more intensely than any other group of pathogens and appear to be well suited for the classical biocontrol approach. They are usually quite specific, easily disseminated and require fewer spores to establish infections than do most other pathogens. Suppression of a weed below economic threshold by a single introduction of the rust fungus into an area where it did not occur naturally would have broad implications. Many weeds have more than one rust disease. Thus there are many other rust fungi that possibly could be used to control such widespread and abundant weeds such as curly dock, water hyacinth, giant rag weed and others.

A rust fungus, *Puccinia chondrillina* was introduced into Australia in 1971 for biocontrol of skeleton weed. It is an autoecious macrocyclic rust. It was a significant factor in suppressing this weed in the Mediterranean region where it is endemic. The pathogen was introduced into Australia and shortly it became widely distributed in denses of skeleton weed. This single introduction has caused spectacular damage to the weed.

Limitations

- Most of the rust diseases rarely build up to lethal levels in nature
- Require dense populations and early infection of their host plants
- Although they do suppress weed growth, it is improbable that rusts or other obligate parasites will damage plants sufficiently to appreciably diminish their numbers.

The classic approach is also being field tested for control of water hyacinth in lakes and water ways in Florida. A newly discovered species of *Cercospora* applied along the edges of infested lakes, spread rapidly throughout the entire mat of floating water surface. This dense mat favours rapid build-up of the fungus. Small lesions on leaf blades and petioles cause leaves to droop into the water and deteriorate. The fungus is sufficiently damaging in natural lakes with low nutrient levels, kills crowns of plants and subsequently causes the plant to 'droop out' or become submerged and deteriorate.

Table 15.1 Biological suppression of weeds with plant pathogens.

Pathogen	Target weed	Crop / Area
(i) Fungi		
Colletotrichum gloeosporiodes(Penz.) f.sp. *aeschynomene*	*Aeschynomene virginica* (Northern Joint Vetch)	Rice
Alternaria macrospora	*Anoda cristata* (Spurred Anoda)	Cotton
Cephalosporium sp.	*Cassia surrantensis* (Brush weed)	Ranges
Puccinia chondrillina	*Chondrilla juncea* (Rust skeleton weed)	Wheat fallow
Cercospora rodmanii *Rhizoctonia solani* *Alternaria eichhorniae* *Uredo eichhorniae* *Myrothecium roridum*	*Eichhornia crassipes* (Water hyacinth)	Aquatic

Table 15.1 Contd...

Pathogen	Target weed	Crop / Area
Ceratocystis fagacearum	*Quercus spp.* (Red Oaks)	Forest
Uromyces rumicis	*Rumex crispus* (Curly dock)	Pastures
Colletotrichum xanthii	*Xanthium spinosum* (Spiny cockle bur)	Soybean
(ii) Bacteria		
Induced mutants of *Pseudomonas solanacearum*	*Solanum dulcamagra* (Bitter night shade)	Tomato
(iii) Nematodes		
Northanguina phyllobia	*Solanum elaeagnifolium* (Silver leaf night shade)	Cotton

15.2 Biological Suppression of Aquatic Weeds

Water, one of the most important natural resources, is often adversely affected by aquatic weeds. Aquatic weeds (floating, submerged and emergent) are those undesirable plants which grow in water resources such as tanks, canals, channels swimming pools, stagnant water spots etc.

15.2.1 Undesirable Effects of Aquatic Weeds

The menace of water weeds is reaching alarming proportions in many parts of the world. The problem is global, but more pronounced in tropical and subtropical zone, where warm weather supports profuse growth of aquatic plants.

- It would be an impediment of water used for agricultural, domestic or recreational purpose
- Increased danger of floods
- Acting as breeding sites for vectors of disease-causing organisms, e.g., mosquitoes which transmit rural filariasis (Elephantiasis)
- Reducing the efficiency of hydro-electric schemes and threatening of native plant communities.

15.3 Aquatic Weeds of Major Concern

(i) Water hyacinth – *Eichhornia crassipes*
(ii) Water- lettuce – *Pistia stratiotes*
(iii) Hydrilla – *Hydrilla verticillata*
(iv) Salvania – *Salvania molesta*
(v) Eurasian Watermilfoil – *Myriophyllum spicatum*
(vi) Alligator weed – *Alternathera philoxerioides*
(vii) Paper bark tree – *Melaleuca quinquinervia*

In addition to the global aquatic weeds, *Salvania natans, Nymphea stellata, Nelumbo nucifera, Vallisnaria spiralis, Typha anqustrata* and species of *Chara* and *Nitella* both algae, are of primary concern in India.

15.4 Biological Control Strategies

In general, there are three approaches for biological control of aquatic weeds (Pieterse, 1990) such as,

(i) use of selective organisms i.e., organisms which attack one or only a few species,
(ii) use of non-selective organisms i.e.. organisms which attack all (or nearly all) weed species present, and
(iii) use of competitive plants i.e., plants which compete with a weed species for one or more critical growth factors.

When selective organisms or plant pathogens are used, biological control can be accomplished by either classical or microbial herbicide strategy. In the classical strategy, which is most suitable for exotic weeds, the pathogens used are mostly rust and other fungi capable of self-dissemination through air-borne spores.

15.5 Aquatic Weeds

Some of the most problematic weeds in India such as water hyacinth, Eurasian watermilfoil and Hydrilla and their biocontrol agents are discussed here under.

15.5.1 Water Hyacinth

It is a large, free floating aquatic weed of the family Pontederiaceae, order Liliales. The plant is a native of the Amazon basin, South America. Introduction of the weed occurred almost simultaneously in the late 19th and early 20th centuries throughout the old world tropics. It now occurs worldwide mainly in the tropical and subtropical regions. It has now attained the status of world's worst aquatic weed (Charudattan et. al. 1990b).

The plant made its entry into Bengal (India) before 1900 and today it occurs throughout the country, except some parts of Rajasthan and hilly regions of northern India and Kashmir. In the Haryana state, it is the number-one troublesome aquatic weed (Aneja, 1992). A total of fourteen virulent fungal pathogens have been recorded on water hyacinth. Of these, *Alternaria eichhorniae, A. alternata, Cercospora piaropi, C. rodmanii* and *Uredo eichhorniae* hold the best promise of biocontrol agent of the weed.

15.5.2 Eurasian Watermilfoil

Eurasian watermilfoil (*Myriophyllum* sp.) is a submerged, perennial aquatic herb that grows as deep as 10 m in fresh to brackish waters. It is monoecious and propagates mainly through autofragmentation, rhizome production and axillary buds. It is native to Eurasia, but has been reported as a weed in several countries in Asia, Africa, Australia, Europe, USA and Canada.

A total of 14 fungal pathogens have been recorded on Eurasian Watermilfoil (Charudattan 1990 a). The type of disease or damage caused due to various pathogens are:

Disease Symptom	*Pathogen*
(a) Systemic stem necrosis	*Acremonium curvulum*
(b) Discolouration and Growth retardation	*Articulospora tetracladia* *Dactylella microaquatica* *Flagellospora stricta*
(c) Die- back	*Fusarium acuminatum, F. oxysporum, F. poae F. roseum F. solani, Sclerotium hydrophyllum*
(d) Stem Necrosis	*Fusarium sporotrichioides*
(e) Shoot lesions	*Mycoleptodiscus terrestris*
(f) Necrosis	*Colletotrichum gloeosporiodes*

15.5.3 Hydrilla

Hydrilla (*Hydrilla verticillata)* is a submerged vascular hydrophyte that is widely distributed in the old world. Tropical Asia is the center of origin. It is an important aquatic weed now widely distributed in Asia, Europe, Australia and the United States of America. Three fungal pathogens namely *Fusarium culmorum, Macrophomina* sp. and *Mycoleptodiscus terrestris* have been suggested as potential biocontrol agents of hydrilla, the number-one submerged weed in the United States.

15.6 Bioherbicidal Strategy

In microbial herbicide strategy, the pathogen is cultured *in vitro* on a large scale and applied in fairly high concentration to the weed, in a manner similar to chemical herbicides. Pathogens used in this strategy can be native or exotic (Templeton 1982, Charudattan 1990c). The concept of mycoherbicide was first introduced by **Daniel** et. al. (1973). It is also called the inundative approach, as pathogens are applied to control weeds within a specific geographical site by an inundative application of inoculum. Inundative application of inoculum of pathogens during early stages of weed growth results in the control of weed infestations without the disease developing beyond the initial lesions into epidemics. The initial lesions caused by the applied inoculum directly cause the death of infected weed seedlings.

Table 15.2 Mycoherbicides available against notorious weeds.

Formulation	Pathogen	Target weed
Devine	*Phytophthora palmivora*	Milk weed vine
Collego	*Colletotrichum gleosporioides f. sp. aeschynomene*	Northern Joint Vetch
Casst	*Alternaria cassiae*	Sickle pod
Biomal	*C. gloeosporioides f. sp. malvae*	Round leaf mallow
Biochon	*Chondrostereum purpureum*	Black berry
Velgo	*Colletotrichum coccodes*	Velvet leaf
Luboa	*C. gloeosporioides f. sp. cuscutae*	Dodder
ABG-5003	*Cercospora rodmanii*	Water hyacinth

CHAPTER - 16

Biological Control of Plant Parasitic Nematodes

Plant parasitic nematodes form an important niche in agro-ecosystems. They feed on plant organs as ectoparasites, semi-endoparasites, and endoparasites and cause reduced plant productivity. They totally depend on their respective hosts for sustenance and survival. They have developed requisite morphological and biochemical adaptations through course of evolution for their feeding habits and mode of survival during host free periods. Endoparasitic nematodes with or without sexual dimorphism constitute the most destructive category of plant parasitic nematodes. They cause diverse damages in agro-ecosystems and yield losses may be enormous. According to an estimate, plant parasitic nematodes cause yield losses of over US $78 billion in agricultural production on world-wide basis. The share of developing countries in this loss is greater than developed countries.

Like many other countries, plant parasitic nematodes are significant impediments in agricultural productivity in India. Root-knot nematodes, reniform nematode, ear-cockle nematode and several ectoparasitic nematodes are distributed throughout the country. Though adequate information from different regions of the country is yet not available or collected, some localized but intensive nematode problems like cereal cyst nematode in Rajasthan on wheat and barley, potato cyst nematodes on potato in the Nilgiri hills of Tamil Nadu, burrowing nematode on banana in Karnataka and Kerala and on coconut, arecanut and black pepper in Kerala are of major concern. Yield loss estimates available in India for a number of crops are not very precise and adequate but available information indicate that they are important in low productivity of a number of crops in different regions of the country.

16.1 Historical Background

Development of two soil fumigants such as DD mixture and EDB in mid-1940s which showed excellent nematode killing properties was the beginning of the efforts for management of plant parasitic nematodes. Christie and Perry (1951) demonstrated ectoparasitism of plants in nematodes by using soil fumigants. In order to increase crop productivity, efforts were made to eliminate nematodes. In that process, development and use of nematicides increased substantially in the following years. Different types of formulations such as granules, volatile liquids, gases, emulsifiable concentrates etc., were manufactured and the nematicides were classified accordingly as soil fumigants and non-fumigants (systemic and non-systemic). They were used for injection in soil drenching, seed treatment, root dip and foliar applications. Most of the nematodes were effective and provided adequate control and a general agreement amongst the nematologists persisted that nematode problems in agriculture can be solved by application of nematicides and the growers became enthusiastic and used over-doses in order to maximize profits. Simultaneously, some non-chemical measures for nematode control were also tried. Such measures include exclusion and avoidance, cropping systems, physical tactics, biological control and deployment of host resistance. Each category of these measures were effective or non-effective on specified crops but essentially all possess their own merits and demerits. Some are still favoured and are constituents of integrated management measures. Even with this parallel development of non-chemical measures, nematode management remained mainly dependent on non-chemical measures. Nematode management remained mainly dependent on nematicides and alternate management strategies were not adequately developed. In 1977, dependence on nematicides received a serious set back. Many of the employees at a chemical plant at Lathrop, California, U.S.A who handled DBCP, a soil fumigant, were found to be either azoospermic or oligozoosperimic. Low sperm counts were found among formulators, custom applicators, farmers and salesmen. The United States protection agency banned production and sale of this nematicide. Several regulatory measures followed as most of the nematicides were found to be the cause of environmental concerns and human health risks.

16.1.1 Adverse Effects of Nematicides

- Nematicides cause soil pollution and adversely affect useful microflora and fauna which interact with pathogens at the soil root interface.

- They are harmful to mycorrhizal fungi, rhizobia, actinorhiza and other microorganisms normally contributing to the nutritional economy of the plants and possibly bring about disturbance in the microbial equilibrium.
- They leave residues in crop plants and soils which present risks to human health and are hazardous to grazing livestock.
- Nematicides pollute water and deteriorate its quality as most of them have been detected in groundwater.

16.2 Symptoms of Damage Due to Nematodes

16.2.1 Root Damage

In search of food, various nematode species inflict visible damage mainly in the form of lesions (necrotizing of cortical tissue) or swellings, cessation of growth and death of root tips or deformations of the root system due to root knots. All these impairments of the feeder roots inhibit their growth and cause them to die. The physiology of the root tissue is also altered in such a way as to generally reduce the functional efficiency of the root. Serious concomitant effects are marked disturbances of water and nutrient uptake by the active root system. The parts of roots attacked by nematodes offer avenues of entrance to soil-borne microorganisms such as fungi and bacteria. These secondary pathogens invade the wounds caused by nematodes or can also infect the weakened plant more easily at different points. Interactions mainly with soil-borne fungi such as *Fusarium* spp., *Rhizoctonia, Verticillium* spp. and *Pythium* spp. greatly intensify the damage already caused by the two groups of pathogens i.e., fungi and bacteria. The necrosis spreads rapidly to the vascular system of the central cylinder of the roots, directly impeding and at an advanced stage completely abolishing the supply of water and nutrients to the plant finally results in complete decay of the affected area of the root.

16.2.2. Shoot Damage

High population densities of nematodes may cause the symptoms due to water stress and reduced nutrient uptake. Aerial symptoms such as wilted, chlorotic and later necrotic foliage leading to death of entire branches, stunting of roots, reduced standing power due to the weakened root system etc. Even if no direct damage to the aerial parts of the plants is yet apparent, substantial yield losses are often still to be expected.

Table 16.1 Nematodes Associated with Crop Plants.

Crop	Nematodes (s)
Rice	*Aphelenchoides besseyi*
Wheat and Barley	*Heterodera avenae*
Maize	*Rotylenchus reniformis*
Finger millet	*Rotylenchus reniformis*
Tobacco	*Meloidogyne incognita*
Pigeonpea	*Heterodera cajani*
Cowpea	*Rotylenchulus reniformis* *Meloidogyne incognita*
Banana	*Radopholus similis*
Citrus	*Tylenchulus semipenetrans*
Grapevine	*Meloidogyne incognita* *M. javanica*
Papaya	*Rotylenchulus reniformis*
Potato	*Globodera rostochinensis* *M. incognita*
Tomato, Brinjal	*Meloidogyne incognita*
Mushrooms	*Aphelenchoides sp.*
Polyhouses	*Meloidogyne spp.* *Rotylenchulus* *Radopholus*

16.3 Techniques Related to Extraction of Nematodes

16.3.1 Nematode Surveys

The best site to sample for nematodes is where feeder roots or feeding sites for nematodes are abundant. In annual field crops, the roots are usually well spread in the 20 cm soil layer but in the case of perennials, 50-100 cm or more and 5 to 30 cm deep are preferred sites. The pattern and intensity of sampling vary with the type of infestation. It may be systematic, random, and stratified random. A composite sample is prepared by combining the cores and sub samples should be drawn after thorough mixing.

The population densities of nematodes and the age structure of their populations vary greatly in time and space. Depending upon the objective of the survey, suitable time for sampling has to be identified. The population density at the time of planting is used for establishing the relationships between nematode density and crop yield.

Depending on the mode of parasitism of the nematode, the sampling of root or plant tissue or soil has to be decided. For ectoparasites viz; *Tylenchorhynchus* sp., *Holplolaimus* sp., *Helicotylenchus* sp., etc. sampling of soil is required. While for endoparasites viz; *Meloidogyne* sp. and

Heterodera sp., observation of root tissues is more reliable. Based on the objective of sampling, type of nematode distribution and precision desired, the size of sample and the number of cores per unit area would depend. The soil sample should be collected in a polyethylene bag and fastened with rubber band with a proper label inside the bag. The label should bear the date, place and the crop. Storage of soil samples usually at low temperature (10-15 °C) is preferable.

16.3.2 Extraction of Nematodes from Soil and Root Samples

There are several techniques such as Baermann funnel method, decanting and sieving, and elutriation methods are available for separation of nematodes from large soil samples. Fenwick can method is generally used for isolation of obese stages of cyst nematodes from soil. For extraction of *Aphelenchoides besseyi*, Baermann funnel technique is suitable .While, the root incubation technique is used for immature stages and males of sedentary parasites.

16.4 Isolation of Some Opportunistic Fungi from Eggmasses of *M. incognita* and Nematode Infested Soil

The infested plant roots and rhizosphere soil are to be collected from different areas where root-knot nematode infestation is prevalent. The eggmasses are picked up with a sterile needle and surface sterilized with 1% sodium hypochlorite for one minute, washed twice in sterile distilled water and placed aseptically on potato dextrose agar medium (PDA) in Petri dishes and incubated at 24 °C for a week for isolation of fungi associated with eggmasses, whereas the soil mycoflora are isolated by serial dilution pour plate technique (Johnson and Curl, 1972) using selective media. The dilution used for isolation of fungi was 10^{-4}. All the fungi that are isolated from eggmasses and rhizospheric soil are identified based on its cultural and morphological characters (Barnett, 1962 and Subramanian, 1971). Axenic cultures of the fungi can be obtained by single spore isolations (Johnston and Booth, 1983) and cultures are maintained on PDA slants.

16.5 Hatching Inhibition Test

For determining the effect of fungal filtrates on larval hatching, cavity blocks are separately pipetted with 5 ml of each dilution of fungal filtrate. Two drops of 0.1 % streptomycin sulphate has to be added to each block to avoid bacterial contamination. Five sterilized healthy eggmasses of nearly uniform size of *M. incognita* are to be transferred to each block. The eggmasses placed in sterilized distilled water serve as control. All cavity blocks should be kept at 25 °C in incubator. Total number of hatched out larvae in each block are counted after 5 days under steriobinocular microscope.

16.6 Juvenile Mortality

An in Vitro test can be carried out with fungal culture filtrates of each of the fungus isolated from rhizosphere soil. For preparation of the fungal filtrates, the fungi are cultured on potato dextrose broth by incubating at 28 °C for 15 days. After incubation, culture filtrate is filtered through whatman filter paper no.1 twice. Filtrates thus obtained are centrifuged at 6000 rpm for 15 minutes and taken as cent percent and designated as standard solution "S". From this, "S/2" and "S/4" are to be prepared by adding required amount of sterilized distilled water. The concen tration grades are kept in cavity blocks @ 5 ml in each block. Two drops of 0.1% streptomycin sulphate are to be added to each block for avoiding bacterial contamination. In each such block 50 juveniles are to be added. Observations on mortality of juveniles can be recorded after 24, 48 and 72 hours.

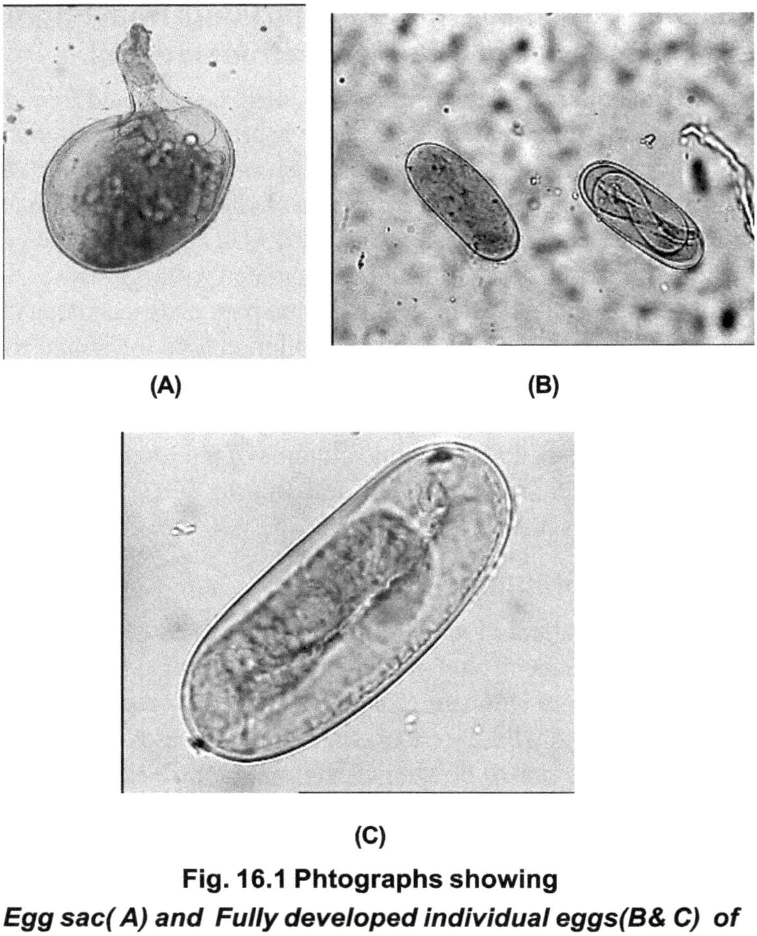

(A) (B)

(C)

Fig. 16.1 Phtographs showing

Egg sac(A) and Fully developed individual eggs(B& C) of M. incognita

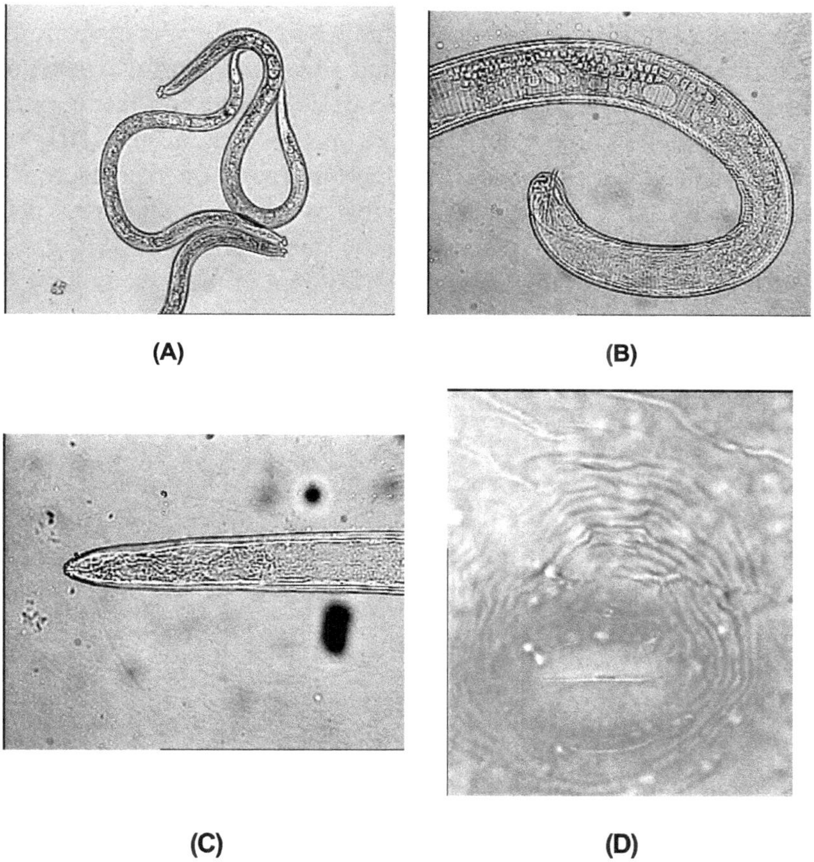

Fig. 16.2 Phtographs showing

Second stage juvenile (A), female (B), male (C) and Perineal pattern (D) in M. incognita

16.7 Bological Suppression of Plant Parasitic Nematodes

Biological suppression of plant parasitic nematodes with fungal and bacterial antagonists is gaining increasing importance as a result of realization that many environmental and health hazards are linked with the use of chemicals. Biological control aims at increasing the parasites, predators and pathogens of nematodes in the soil, in order to increase the mortality of plant parasitic nematodes. This can be done by introducing the biocontrol agents directly which would enhance the biotic component of environmental resistance.

16.7.1 Types of Biocontrol Agents

Fungi, bacteria, algae, predaceous nematodes, viruses, tardigrades, earthworms, mites and insects have been reported to be associated with nematodes and most of them have been proved to be important biocontrol agents. Among them, fungi and bacteria have been studied in detail. Apart from direct colonization they also produce metabolites which are harmful to nematodes. Fungi possessing the capacity to destroy or deleteriously affect nematodes vary considerably in both biology and taxonomic relationship. They range from obligate endoparasitic, many of which are zoosporic to predaceous trap forming species and opportunistic fungi, that colonize reproductive structures such as cysts and eggs. Endoparasitic fungi are *Acrostallagmyces, Harposporium, Meria* and *Catenaria.* Very few of these fungi have been considered as potentially useful biocontrol agents because they are not cultivable and they are unable to parasitize the stylet forming phyto-nematodes because the spores cannot be ingested (Cooke, 1968).

16.7.2 Predaceous Fungi

They are also known as nematode trapping fungi. They consist of sparse mycelium which has been modified to form organs capable of capturing nematodes. They are grouped into sticky and mechanical traps.

Sticky traps are of three types :

- Sticky branches, e.g. *Dactylella lobata*
- Sticky net work, e.g. *Arthrobotrys oligospora*
- Sticky knobs, e.g. *Dactylella ellipsospora*

While mechanical traps are of two types

- Non-constricting ring. e.g. *Dactylaria candida*
- Constricting ring, e.g. *Dactylella bambicodes*

16.7.3 Opportunistic Fungi

These fungi are capable of colonizing nematode reproductive structures and deleteriously affecting them. Nematodes belonging to *Globodera, Heterodera* and *Meloidogyne* at sedentary stage of their life cycle are vulnerable to attack by the fungi either within the host plant roots or exposed to root surface or within soil. They also attack eggs. As soon as these fungi contact eggs they rapidly grow and colonize eggs that are in the early embryonic developmental stage. They lose their parasitic activity if the juvenile is formed within the eggs. The differential vulnerability of egg and larva is due to chitinolytic activity of these fungi. Egg shells possess chitin but larval cuticle lacks it which limits the fungus to parasitize the egg shell. The main types of destructive activity are thought to be enzymatic disruption of nematode structural elements such as egg shell and larval cuticle and physiological disturbances brought about by biosynthesis of diffusible toxic metabolites

(Morgan Jones and Rodriguez - Kabana, 1985). General symptoms encountered due to fungal infection are granular shriveled eggs, black eggs containing spore like bodies and eggs with contorted juveniles.

Several fungi such as *Paecilomyces lilacinus, Verticillium chlamydosporium, Trichoderma harzianum, T. viride* and *Gliocladium virens* have been reported to parasitize eggs of plant parasitic nematodes.

16.7.3.1 Paecilomyces lilacinus

Paecilomyces lilacinus has been one of the principal genera of interest in biological control studies in recent years. It is a species with a cosmopolitan distribution particularly in warmer regions of the world and it was first observed by Lysek (1976) found in association with eggs. It is by far the most promising and practicable biocontrol agent for the management of root-knot and cyst nematodes.

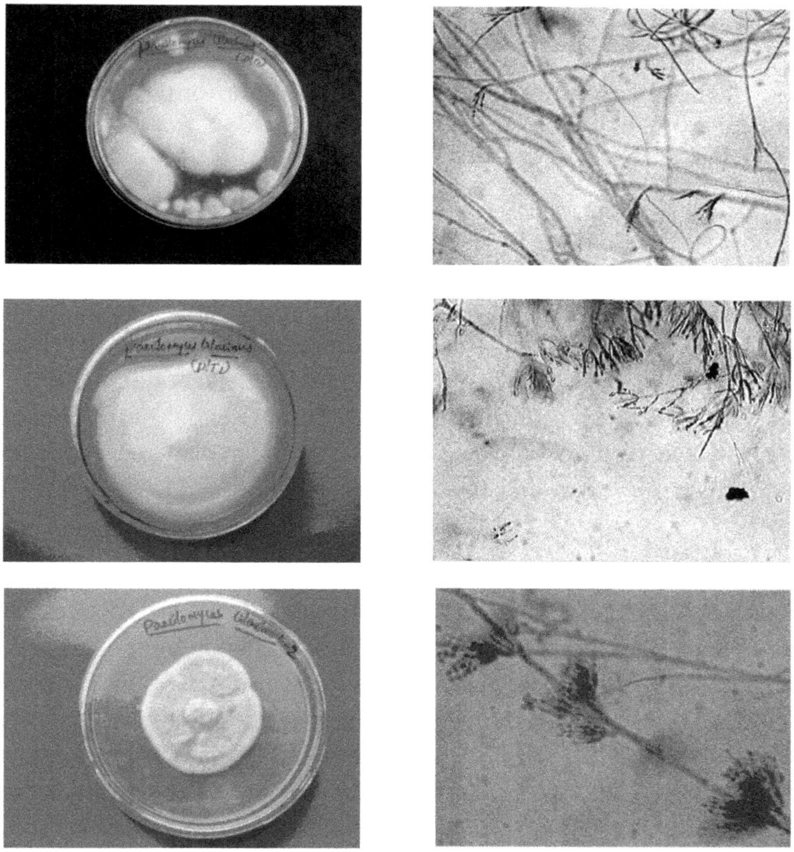

Fig. 16.3 Contd...

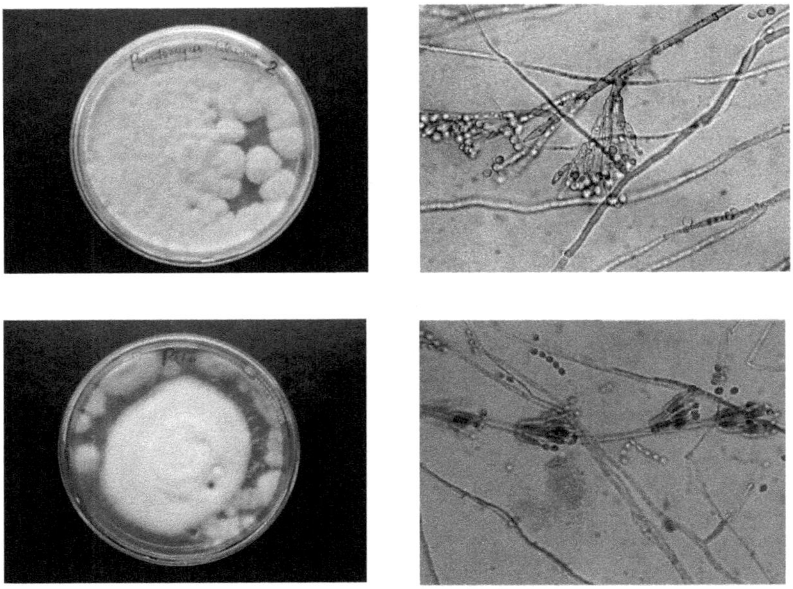

Fig. 16.3 Pure cultures and corresponding photomicrographs of different isolates of (5 isolate) *Paecilomyces lilacinus* isolated from different areas of Rayalaseema districts

Mode of action

Morgan – Jones et. al. (1984) showed that hyphae of *P. lilacinus* readily penetrate egg shells of *M. arenaria* through small pores dissolved in the vitelline layer. The effect seems to involve disruption of embryonic development and death of larvae, resulting in the reduction of nematode populations. This fungus was also capable of invading mature females and cysts of *Globodera pallida* (Jatala et. al. 1979) and also effective for the management of reniform nematode infecting tomato and brinjal (Parvatha Reddy and Khan, 1988, 1989).

Nahat et. al.(1995) reported the application of *P. lilacinus* at the rate of 8 g /kg soil as the best treatment for the management of root-knot nematode in betelvine. This fungus was also used for management of *Radopholus similis* in betelvine. Application of fungi before nematode inoculation was found to be very effective against *R. similis*. Here reduction in nematode population is due to rapid colonization of root tissues by the fungus thereby reducing the feeding of *R. similis* (Sosamma et al. 1994). The most important finding is that *P. lilacinus* did not cause any adverse effect on black pepper seedlings and found to survive and remain effective even after two years of inoculation in greenhouse conditions (Sosamma and Kosby, 1995).

16.7.3.2 Verticillium chlamydosporium

Verticillium spp. are another important biocontrol agent of nematode which had received attention in recent years. Among different species, *V. chlamydosporium* has been recorded in cyst and / or eggs of *Globodera, Heterodera* and *Meloidogyne*. For the first time in India, this fungus was isolated from *Trophotylenchulus piperis* infesting black pepper (Anon. 1995). This fungus was proved to be effective in suppressing root-knot nematode in cardamom and black pepper. *V. chlamydosporium* readily penetrated the root-knot egg masses but no mycelial penetration was observed in the case of juveniles of *M. incognita* and *R. similis* (Sreeja et al. 1996). Morgan – Jones et. al. (1983) reported that *V. chlamydosporium* has been shown to be capable of preventing egg hatching of *M. arenaria* and to colonize eggs by hyphal penetration.

16.7.3.3 *Trichoderma* spp.

Different species of fungi viz., *Trichoderma viride, T. harzianum, T. hamatum, T. aureoviride, T. polysporum, T. longibrachiatum, T. koningii, T. pseudokoningii* and *Gliocladium virens* were found to colonize eggmasses of *M. incognita* but the colonization level varied between isolates. Among these, *T. harzianum, T. longibrachiatum, T. koningii, T. viride* and *G. virens* were the best colonizers. Distortion of eggs was commonly seen and was more pronounced with regard to *T. harzianum T. viride* and *G. virens* suggesting the involvement of toxic metabolites. Culture filtrate of all the fungi killed the second stage juveniles of *M. incognita* in 24 hours. In pot culture studies also one isolate of *T. harzianum* suppressed root-knot nematode in cardamom (Anonymous 1995).

16.7.3.4. Vesicular – Arbuscular **Mycorrhizae** *(VAM)*

There are a group of obligatory symbiotic endophytes in the family Endogonaceae which colonize plant roots and project an extensive network of hyphae into the surrounding soil. The hyphal net work increases the absorption area of the root system and therefore aids in water uptake, absorption and translocation of nutrients particularly phosphorus. Hussey and Rancodory (1982) and Smith (1987) listed a number of possible hypotheses to explain the beneficial effect of VAM fungi on plants parasitized by nematodes.

- VAM fungi enhances the uptake of phosphorus and other nutrients, thereby improving plant vigour and growth which will offset the yield loss caused by nematodes.

- It will reduce the root exudates or alter physiologically responsible for stimulating egg hatching or the chemotactic alteration of nematodes to roots.
- Parasitization of female nematodes and their eggs.
- Retardation of nematode development or reproduction within or near root tissue either by the production of nematostatic compounds or by competition for space or host photosynthates.

Mycorrhizal fungi have shown antagonistic influence on the population of plant parasitic nematodes which may be due to increased aminoacid contents, especially arginine in the mycorrhizal roots (Nemee and Meridith, 1981) or due to the presence of additional amount of phosphorus in roots. VAM can markedly alter the plant response to nematodes. For example, tomato roots colonized by *Glomus fasciculatum* and infected by *M. incognita* or *M. hapla* exhibited fewer and smaller galls than did nematode infected non-mycorrhizal plants (Bagyaraj et al. 1979). Small giant cells with few nuclei and retarded nematode development are observed in mycorrhizal tomato roots infected with *M. incognita* (Sikora, 1979). In addition to adverse effect on nematode penetration and development, there are indications that nematode reproduction is also suppressed on mycorrhizal plants.

Glomus mosseae was the most effective VAM fungi in reducing *M. incognita* population in roots and promoting growth of crossandra (Nagesh, 1996b). The colonization of roots by VAM fungi may prevent nematode penetration and their establishment mechanically increased nutritional status, increased sugar content of roots and increased levels of phosphorus in roots or increased host tolerance.

Most of the research conducted under greenhouse conditions indicated that mycorrhizal fungi have the potential to nullify the damage due to phytonematodes but intensive investigations are need to be carried out under field conditions to assess their potential utilization as biocontrol agents. Since VAM fungi are obligate parasites, they cannot be cultured on large scale which hinders their effective utilization in integrated nematode management programmes under field conditions.

16.7.4 Bacteria

Bacteria may be categorized into two groups based on their mode of action.

(a) Bacteria which release metabolites that have a killing or inhibitory effect on plant nematodes E.g. *Bacillus, Azotobacter.*

(b) Bacteria which parasitize directly on nematodes, thus affecting the penetration and reproduction of nematodes Eg: *Pasteuria.*

16.7.4.1 *Pasteuria penetrans*

It is an obligate parasite of nematodes that has been gaining importance as a biocontrol agent. This was first described by Thorne in 1940 as a sporozoan, *Dubosquia penetrans,* later renamed as *Pasteuria penetrans* by Sayre and Starr (1985). It is known to parasitize 205 nematode species belonging to 96 genera from 10 orders of the phylum Nematoda. It is a mycelial endospore forming bacterium with a life cycle remarkably synchronizing with the development of its host.

16.7.4.2 *Life Cycle* of *Pasteuria penetrans*

The bacterial spores present in the soil adhere to the cuticle of the second stage juveniles as they migrate through the soil. The spore encumbered nematodes enter into the root and begin to feed. Approximately 30% of the spores attached to the cuticle germinate (Stirling, 1984). The germ tube develops into a vegetative spherical colony consisting of septate mycelium which eventually fragment to daughter colonies and fills the body cavity of developing females. Upon decomposition of dead females, spores are liberated in the soil and remain there for long periods of time. Infected adult nematodes are filled with as many as two million spores.

Pasteuria penetrans has been demonstrated to exert effective biocontrol of root-knot nematode on tomato (Mankau, 1975). There has been a significant reduction in the root gall index and the number of eggmasses of *M. incognita* by about 52 and 60 per cent, respectively due to *P. penetrans* antagonism in brinjal (Nagesh et. al. 1996). In India, two different strains of *P. penetrans* have been reported i.e., one infecting *M. incognita* but not *Heterodera cajani* and the other infecting *H. cajani* but not *M. incognita* (Singh and Dhawan, 1990).

Advantages

- Resistant to heat and desiccation
- Compatible with nematicides
- Ability to survive for more than two years in soil
- It reduces root penetration by nematode juveniles
- Inhibits egg formation, reduces nematode population in soil

Because of the above characteristics, it qualifies as the most potential bacterial biocontrol candidate which could be used on a large scale under field conditions.

Limitations

- Lack of mobility
- Dependence on dispersal agents

- Host specificity
- Because of its obligate nature of organism, large scale production would be rather difficult.

16.7.5 Actinomycetes

These are commonly distributed in most of the cultivated soils. The **'avermectins'** are a new class of macro-cyclic lactones isolated from the soil actinomycetes, *Streptomyces avermitilis*. These compounds are reported to be possessing insecticidal, acaricidal and nematicidal properties (Putter et. al.1981).

The available literature on avermectins indicates that *M. incognita, M.javanica, M. arenaria* and *Tylenchulus semipenetrans* are the commonly used test organisms to assess the nematicidal activity of these metabolites on tomato and citrus (Garabedian and Van Gundy, 1983). In citrus, a monthly rate of 1.1 kg a.i. per ha of avermectins for 7 months gave maximum increase in yield and reduction of *T. semipenetrans* population. While, avermectins B1 and B2 have shown high toxicity against *M. incognita* in greenhouse conditions when incorporated into soil, avermectins B2a was found to be more potent than B1 (Putter et. al.1981).

16.8 Ideal Characteristics of Successful Biocontrol Agent (Jatala, 1986)

- It must not be pathogenic to plants, humans or other animals.
- It must reduce or suppress nematode populations in large quantities so that after a given period the nematode population gets reduced to a level below the damage threshold and reaches an equilibrium.
- It must be capable of adapting or tolerating varying soil environment and structural conditions.
- It should be biologically competitive with other existing or introduced parasites or predators in the host ecosystem.
- It must have a high reproductive potential so that sufficient ineffective propagation units are produced to maintain a high population.
- The persistence and survival capacity of the organism or its ineffective structures must be high in soil environmental extremes and in the absence of the host.
- The propagules or infective units should be able to germinate and grow rapidly so that they are effective in invading and killing nematodes prior to the completion of the developmental and reproductive cycles.
- It should have the capacity to produce antibiotics or other deleterious compounds that either inhibit nematodes or other antagonists to increase its survival.

- It should be capable of affecting a diverse number of nematode species.
- It should be capable of disseminating readily in the soil where it is applied.
- It can be readily and inexpensively mass produced.
- It should have the characteristic of withstanding long-term storage until ready to be applied.
- In addition to suppressing nematode population, it may also suppress the populations of other pathogens, pests and parasites of important crops.
- It should not get suppressed by the application of fertilizers, pesticides and other agricultural practices.
- It should have certain genetic stability in order to ensure long-term effectiveness.
- The application of a successful biocontrol agent to field soils should result in increased crop yields and thus become economically viable.

16.9 Integrated Nematode Management (INM)

Based on the principles of prevention, population reduction and tolerance, integrated nematode management (INM) procedures are implemented to alleviate the nematode problems. The major objective of INM is to stabilize populations of target nematodes at acceptable levels resulting in favourable long-term socio-economic and environmental consequences. In practice, integration of different methods of nematode control is the rule than the exception. In most cases, practical control of a nematode disease involves integration of several diverse control measures. In this context, biological control agents form an ideal component in the INM along with botanicals and, to a limited extent, chemicals.

16.9.1 Mixture of Antagonists

Integration of two bioagents, *P. lilacinus* and *V. chlamydosporium* resulted in combined and complementary effects for the successful management of *M. incognita* infecting tomato (Rao and Parvatha Reddy, 1992a).

Nagesh et. al. (1995) reported that combined application of *P. lilacinus* and *Pasteuria penetrans* enhanced plant growth and flower yield of crossandra besides reducing root galling due to *M. incognita*. Integration of *Glomus deserticola* with *V. chlamydosporium* gave effective management of *M. incognita* on tomato seedlings and increased the egg parasitization by the bioagents (Rao et. al. 1997 a).

The population of *M. incognita* and *M.javanica* were suppressed effectively and yields of tomato were greater when *P. penetrans* and *P. lilacinus* were applied together (Maheswari and Mani, (1988). When black pepper vines were combinedly inoculated with *P .penetrans* and *P. lilacinus,* they had put out 23-112 percent increase in plant growth over control and were effective in the management of root-knot nematodes (Sosamma and Koshy, 1995).

16.9.2 Combination of Bioagents and Botanicals

Application of neem cake and castor cake both at 10 g per plant along with an antagonistic fungus, *Paecilomyces lilacinus* at 4 g (grown on paddy seeds) per plant was found to be extremely effective in the management of citrus nematode, *Tylenchulus semipenetrans.* It reduces the citrus nematode population with a consequent growth of acid lime seedlings (Parvatha Reddy and Khan, 1991).

The parasitization of citrus nematode females with *T. harzianum* increased in the presence of the oil cakes. *T. harzianum* in combination with neem, karanj and castor oil cakes was effective in increasing the growth of acid lime seedlings and reducing the citrus nematode population both in soil and roots. Integration of neem cake at 400 g per plant with *P. penetrans* at 100 g soil (300 spores) was found effective in reducing *Radopholus similis* population both in soil and roots of banana by more than 50 percent and increased plant growth parameters (Channabasappa et. al. 1995).

Incorporation of neem cake / castor cake / neem leaf / castor leaf / pongamia leaf in nursery beds at 400 g /m^2 and application of spore suspension of *P. lilacinus* or *V. chlamydosporium* in tomato and brinjal nursery beds amended with the above plant products facilitated the effective management of root-knot nematode, *M. incognita* and reniform nematode, *R. reniformis.* Further, root dip treatment of seedlings in 5% aqueous suspension of neem leaf cake mixed with the spores of *P. lilacinus* or *V. chlamydosporium* for 20-30 minutes before transplanting was effective in the management of root-knot and reniform nematodes in the main field. (Rao and Parvatha Reddy, 1996).

16.9.3 Combination of Bioagents and Endomycorrhizae

Interestingly, integration of bioagent, *P. lilacinus* and endomycorrhizae, *Glomus mosseae / G. fasciculatum* have culminated into the successful management of *M. incognita* infecting tomato. This phenomenon facilitated standardization of a strategy wherein inoculation of mycorrhizae and bioagent in the nursery beds protected the seedlings of tomato from the attack of *M. incognita.* Further, these mycorrhizal seedlings, colonized either with

G. mosseae or *G. fasciculatum* can be given a root-dip treatment with spore suspension of *P. lilacinus* of these nematodes in the main field after transplanting (Rao et. al. 1993 b,d).

16.9.4 Combination of Bioagents, Endomycorrhizae and Botanicals

After investigating the interactive effects of useful components of management such as bioagents, botanicals and endomycorrhizae, strategies to integrate these components for the management of root-knot and reniform nematodes infecting tomato and brinjal have been developed. These strategies include inoculation of *G. fasciculatum* in the castor cake amended nursery beds followed by the root-dip treatment of mycorrhizal seedlings of brinjal in spore suspension of *P. lilacinus* for the management of *R. reniformis*. Similarly, inoculation of *G. mosseae* in bean leaf / neem cake amended nursery beds followed by the root-dip treatment of mycorrhizal seedlings of tomato in spore suspension of *P. lilacinus* was effective for the management of *M. incognita* under field conditions.

Integration of a bioagent *Verticillium lecanii,* endomycorrhiza *Glomus mosseae* with botanicals improved the growth of crossandra and reduced the population of *M. incognita. Glomus mosseae* reduced the requirement of phosphatic fertilizer and favoured the antagonistic potential of *V. lecanii* against *M. incognita.*

16.9.5 Combination of Endomycorrhizae and Botanicals

Inoculation of endomycorrhiza, *Glomus fasciculatum* in the soils amended with neem cake was found to be effective in the management of *T. semipenetrans* on acid lime. This strategy can help in combating the menace of citrus nematode at nursery stage and also provide highly mycorrhizal seedlings of acid lime for transplanting in the main field for the management of *T. semipenetrans* under field conditions (Parvatha Reddy et. al. 1995). Integration of karanj cake with *G. mosseae* gave maximum banana fruit yield, while neem cake + *G. mosseae* was most effective in reducing *Radopholus similis* population both in soil and roots (Parvatha Reddy et. al. 1997).

Inoculation of *G. mosseae* or *G .fasciculatum* in the nursery beds amended with neem cake / castor cake / neem leaf / calotropis leaf reduced the infestation of root-knot and reniform nematodes to the maximum extent on tomato. Amendment of botanicals in the nursery beds increased the multiplication of endomycorrhizae and colonization of tomato root, which, in turn, could protect the crop from these nematodes in the main field resulting in increased yields (Rao et. al. 1996, 1997).

Similarly, inoculation of *G. mosseae / G. fasciculatum* in the brinjal nursery beds and subsequent application of 5% aqueous extracts of neem cake / castor cake/ neem leaf / calotropis leaf in the nursery beds resulted in the effective management of root-knot and reniform nematodes and yielded healthy brinjal seedlings which would withstand the attack of these nematodes after transplanting in the main field (Rao et. al. 1993 a, c).

The root colonization of VAM fungi increased significantly in the presence of neem cake, which in turn, improved their efficacy in reducing *M. incognita* population in crossandra roots (Nagesh and Parvatha Reddy, 1996b).

16.9.6 Combination of Bioagents and Physical Methods

Soil solarization of nursery beds and application of bioagents such as *P. lilacinus* or *Trichoderma* spp. improved growth of cardamom seedlings by suppressing root-knot nematode population (Eapen and Venugopal, 1995).

16.9.7 Combination of Bioagents and Chemicals

An integrated management of the citrus nematode, *T. semipenetrans* infecting acid lime based on application of the parasitic fungus, *P .lilacinus* at 4 g per plant which is grown on paddy seeds along with carbofuran at 1 kg / ha has been developed. It is found to substantially reduce the nematode population both in soil and roots and to increase plant growth parameters (Parvatha Reddy et. al. 1996a).

Integration of bioagents, *P. lilacinus* with carbofuran at 1 kg ai / ha was found effective in the management of reniform nematode *R. reniformis* infecting tomato and brinjal. Integrated management of root-knot nematode infecting crossandra was obtained by rational combination of biocontrol agents *V. chlamydosporium / Trichoderma harzianum* with aldicarb at 0.05 kg ai/ ha (Parvatha Reddy and Nagesh, 1995).

Maheswari et. al. (1987) reported that application of *P. penetrans* in combination of carbofuran, aldicarb, miral, sebufos and phorate significantly improved plant growth of tomato by greatly reducing galling due to *M.javanica.* Significant reduction of *M. incognita* population in soil and root-knot index and increased plant growth parameters of betelvine reported due to the application of *P. lilacinus* at 4 g per plant and carbofuran 0.75 a per ha (Nakat et. al. 1995).

The INM strategies developed in the above manner are eco-friendly and aim at efficient utilization of promising bioagents, botanicals, chemicals, physical methods and endomycorrhizae for obtaining high yields by suppressing parasitic nematodes.

Pure cultures and corresponding photomicrographs of some of antagonistic fungi isolated from soil and egg masses of *M. incognita.*

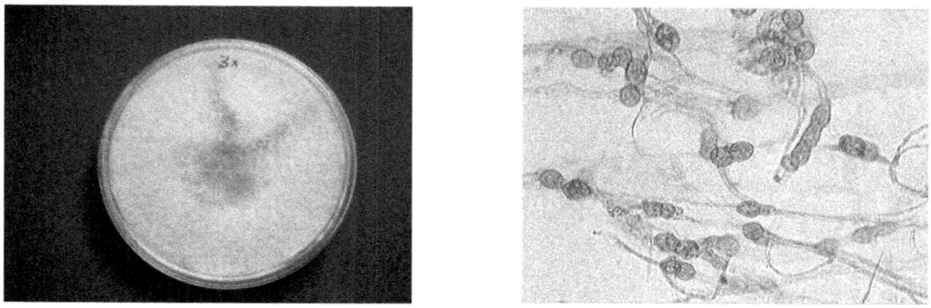

(A)

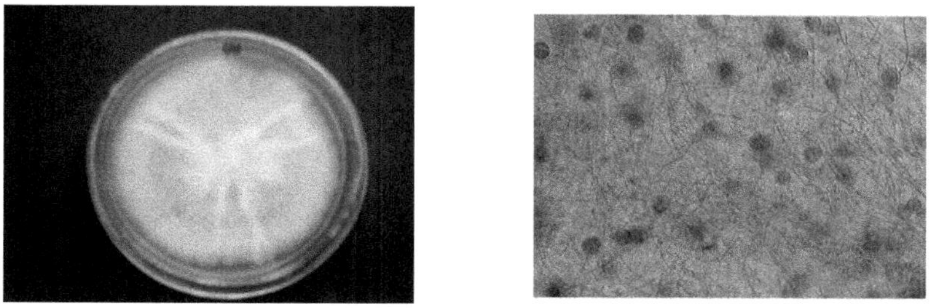

(B)

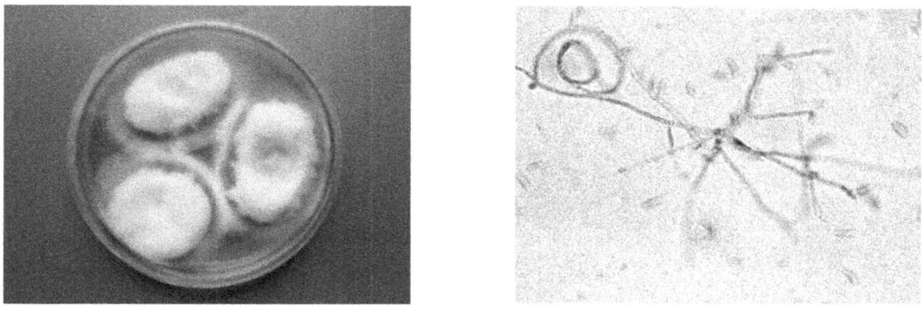

(C)

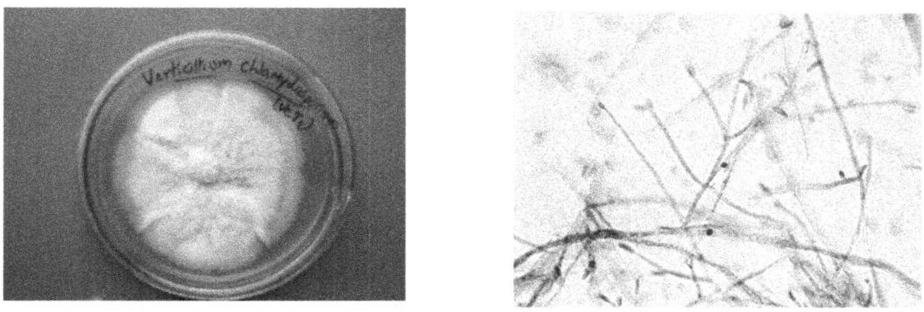

(D)

Fig. 16.4 Phtohraphs showing

Pure cultures and corresponding photomicrographs of different isolates of *Verticillium chlamydosporium* (A, B, C D).

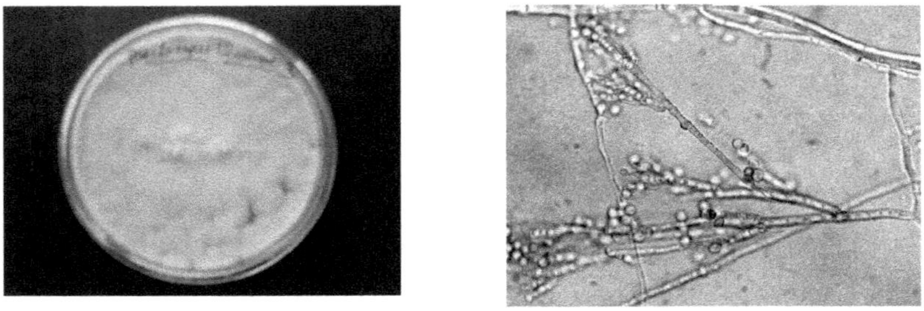

(E)

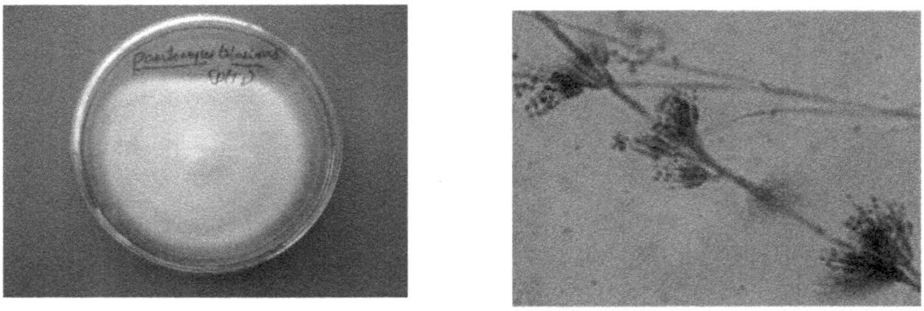

(F)

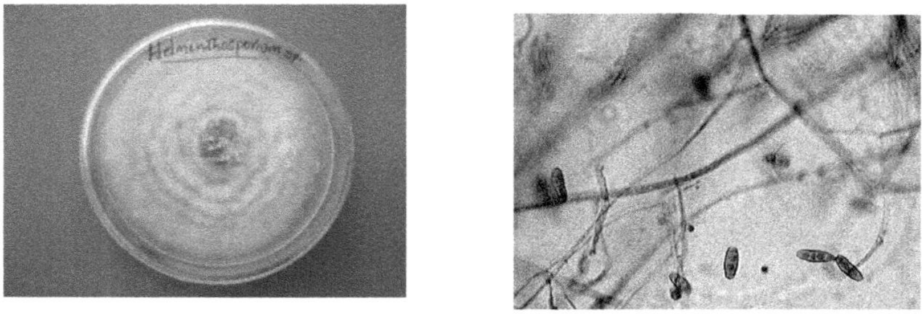

(G)

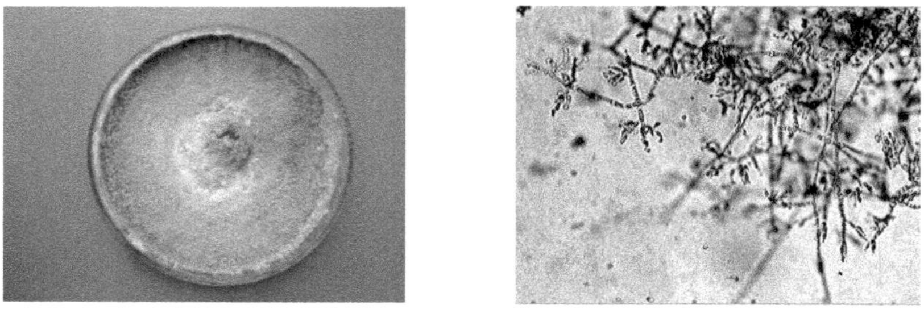

(H)

Fig. 16.5 Phtohraphs showing

Pure cultures and corresponding photomicrographs *Paecilomyces lilacinus* (E & F) *Helminthosporium* sp.(G), *Trichoderma harzianum(H)*.

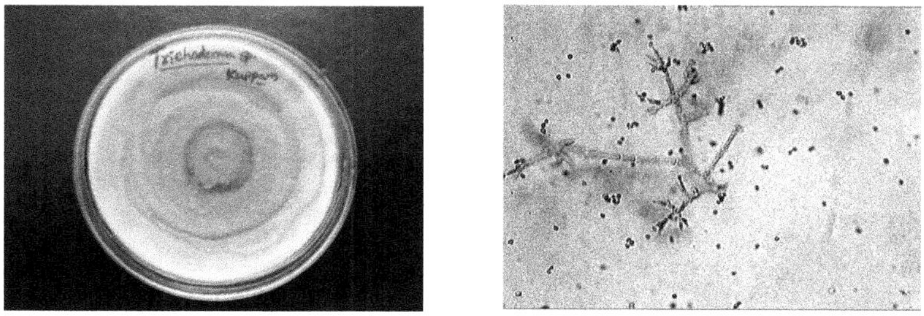

(I)

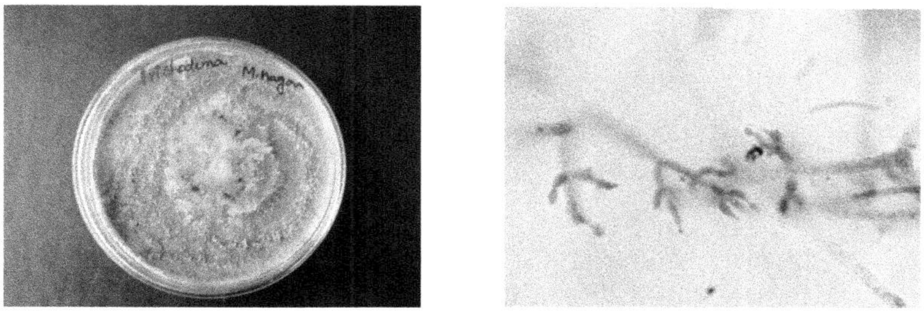

(J)

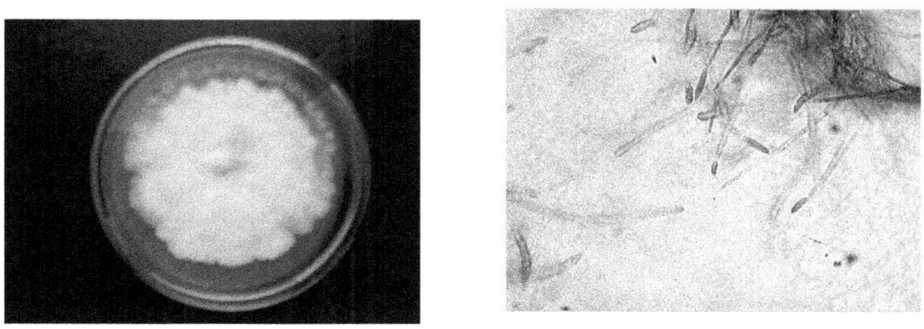

(K)

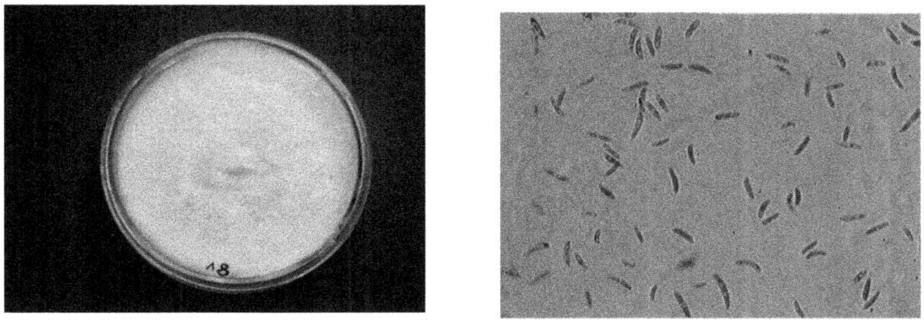

(L)

Fig. 16.6 Phtohraphs showing

Pure cultures and photomicrographs of *Trichoderma viride* (I & J), *Fusarium solani(K)*, *F .oxysporum (L)*.

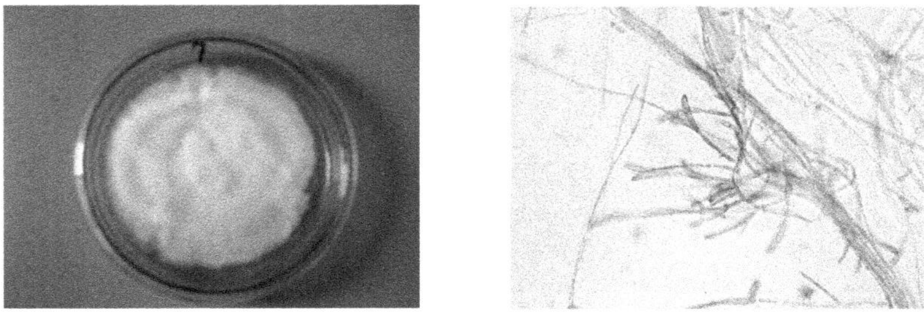

(M)

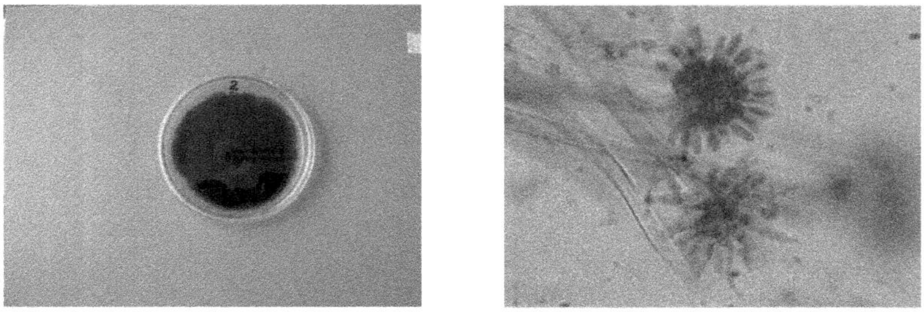

(N)

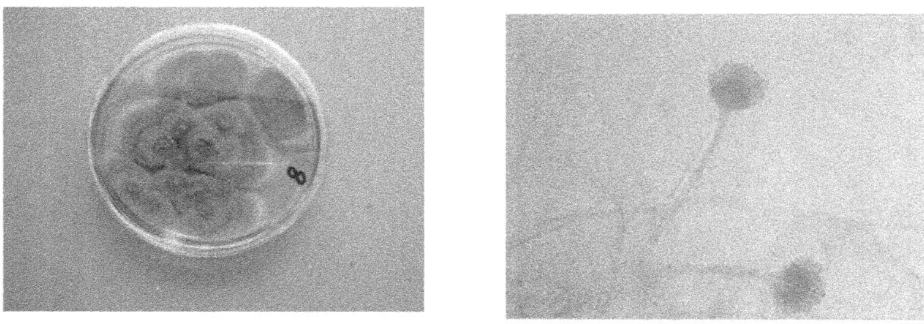

(O)

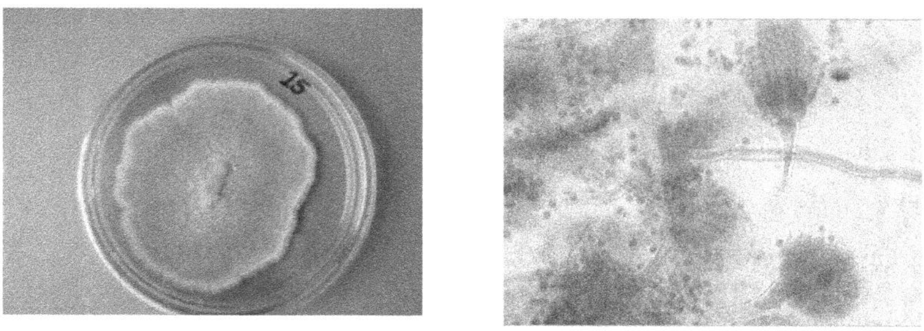

(P)

Fig. 16.7 Phtohraphs showing

Pure cultures and photomicrographs of *F. sporotrichioides* (M), *Aspergillus niger* (N), *A.flavus* (O), *Aspergillus* sp. (P).

Nematode damage symptoms in important Crops of Andhra Pradesh

Above ground symptoms

Uneven crop stand of tomato

Chlorosis in chillies

Decline of pomegranate

Below ground symptoms

Galls formed by root knot nematode, *Meloidogyne incognita*

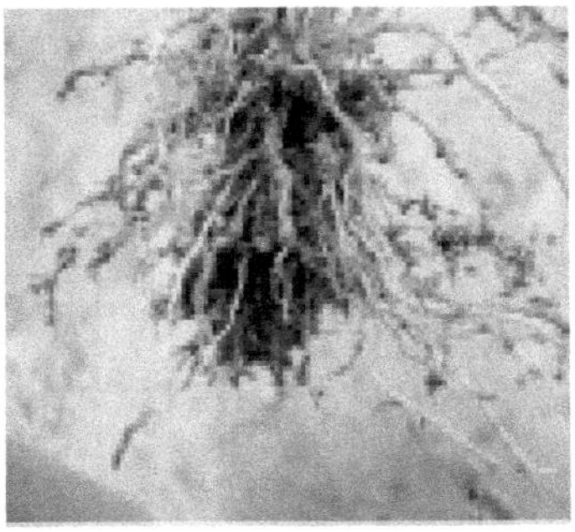

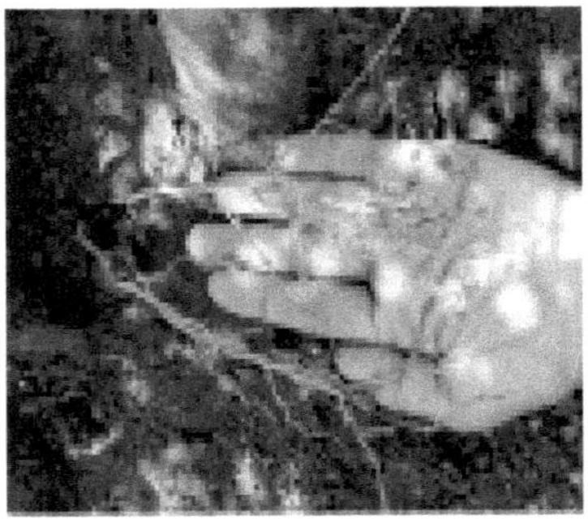

Galls in root system caused by RKN

Above ground symptoms

Slow decline of citrus

Stunting and chlorosis of Crossandra

Patchy growth of tuberose

Below ground symptoms

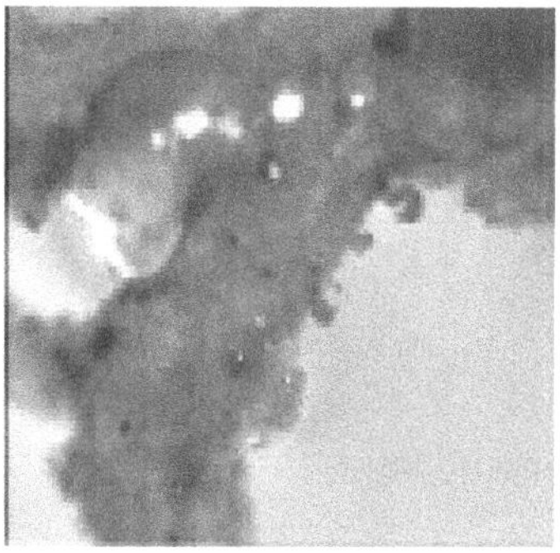

Tylenchulus semipenetrans in roots

Root lesions caused by *Pratylenchus coffeae*

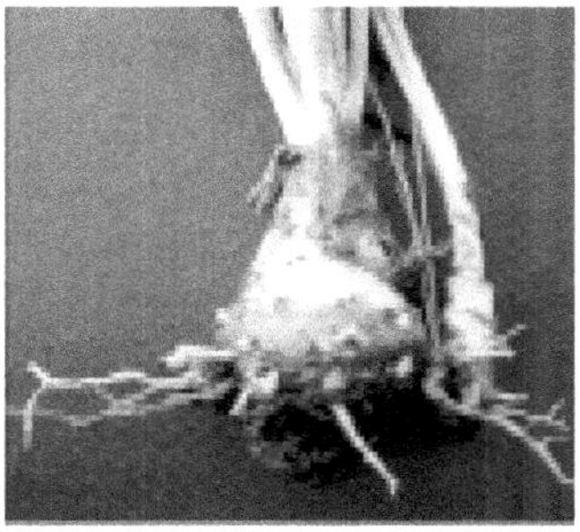

Galls in feeder roots

Above ground symptoms

Chlorosis caused by cyst nematode

Poor vine growth and Chlorosis

Stunted plants with greener foliage

Below ground symptom

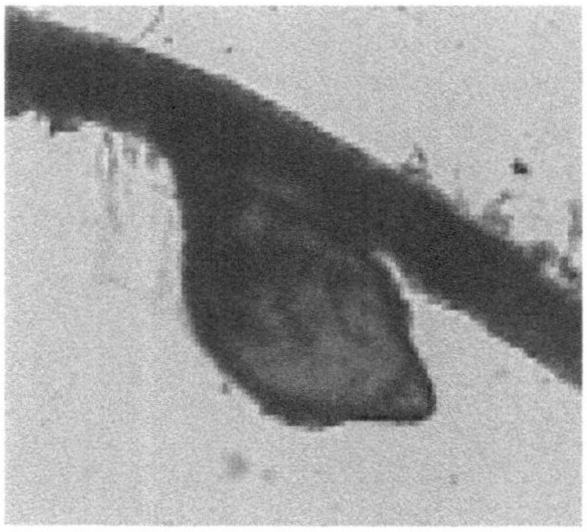

Heterodera cajani in root system

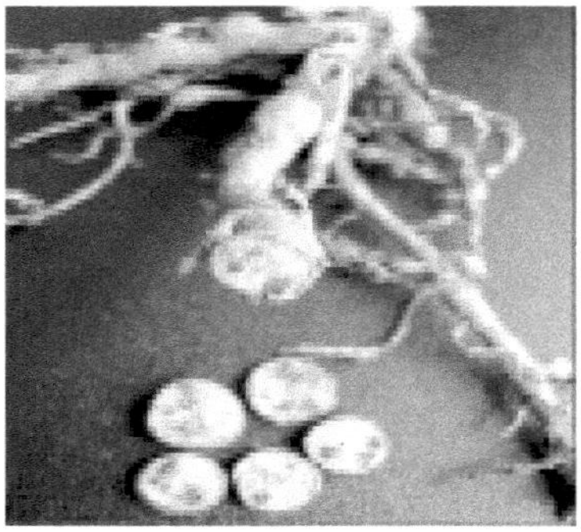

Females of RKN embedded in root galls

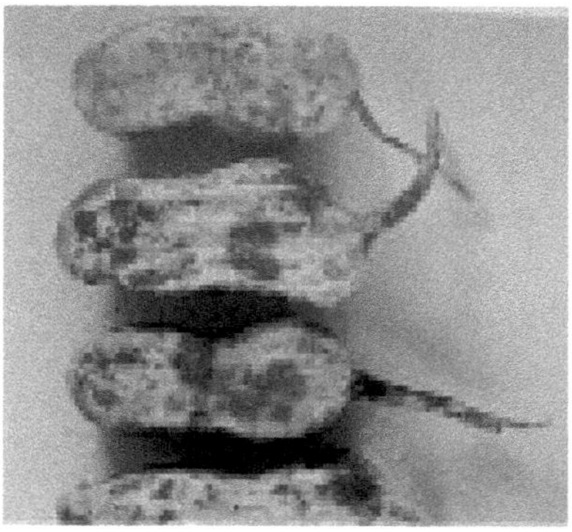

Lesions in pods caused by *Tylenchorhynchus brevilineatus*

Above ground symptoms

Poor crop stand in Castor

Poor seedling establishment

Chlorosis of Rice plants

Below ground symptoms

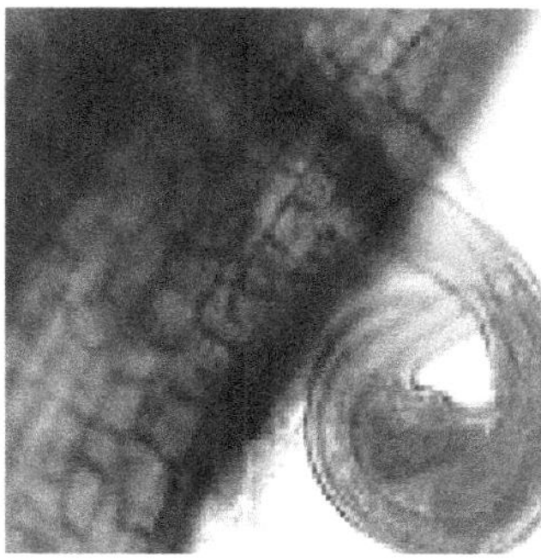

Rotylenchulus reniformis in root system

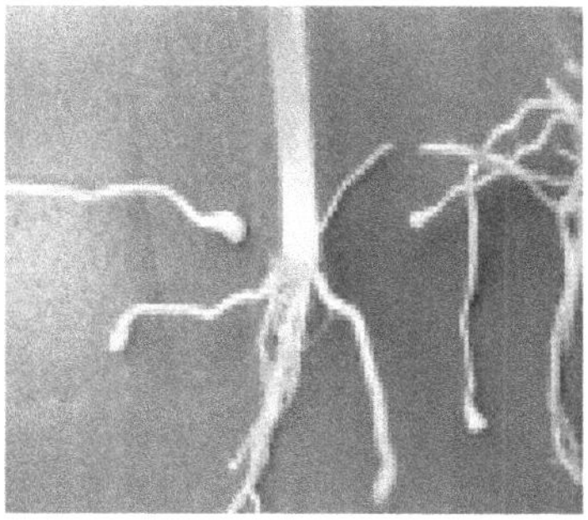

Galls caused by *M.graminicola*

Root lesions caused by *Hrishmanniella oryzae*

Chapter - 17

Registration Requirements and Quality Parameters for Antagonistic Organisms

Chemical pesticides have been used traditionally for the control of pests as these have been effective, dependable and economical. However, several problems such as development of resistance in pests to pesticides, pest resurgence, toxic residues in food, water and soil, destruction of natural enemies and disruption of ecosystem has been resulted. In view of this, biopesticides are gaining importance for the control of pests in agriculture, forestry, horticulture and public health programmes. Microbial pesticides are the biopesticide formulations containing bacteria, fungi viruses, nematodes, protozoa etc. They also include antagonistic organisms for the biological control of plant diseases, whereas, botanical pesticides are those which are based on plant materials such as neem etc.

Increased emphasis is being given by government agencies and biopesticide industry to promote their use under ambit of Integrated Pest Management (IPM) to reduce dependence on hazardous chemical pesticides under the IPM programmes, biopesticides being cost-effective and environment-friendly are the best alternative to manage pests and diseases.

There are a number of antagonistic fungi and bacteria which are used in the biological control of plant diseases. The import, manufacture, sale, transport, distribution and use of pesticides are regulated in India under the Insecticides Act, 1968 and Rules framed thereunder. For this purpose, the pesticides should be included in the 'Schedule' to the Insecticides Act. The antagonistic organisms included in the Schedule of Insecticides Act to regulate their manufacture and use in the country by the Central Insecticides Act, 1968 constituted under the Insecticides Act, 1968 are, *Bacillus subtilis, Gliocladium* spp. *Pseudomonas* spp. and *Trichoderma* spp. The guidelines for inclusion of new chemicals and substances on the Schedule to the Insecticides Act are given in the Annexure I.

17.1 Data Requirements for Registration

The registration committee (RC) constituted under the Insecticides Act, 1968 normally allows commercialization in terms of import, manufacture, sale and use of pesticides only after the grant of regular registration under the Act. The data requirements for grant of certificate of registration under the Insecticides Act, 1968 in respect of antagonistic organisms such as *Trichoderma, Bacillus subtilis, Pseudomonas, Gliocladium* under section 9(3B) and 9(3) are detailed in Annexure –II, III and IV respectively. Some of the microbial pesticides testing laboratories identified by the expert group are given in Annexure – VI.

Guidelines For Inclusion of New Chemicals / Substances in the Schedule To The Insecticides Act, 1968 As Approved By The Central Insecticides Board in its 22nd Meeting.

1. Common Name :
2. Chemical Name:
3. Literature Support:
   (i) Chemistry
   (ii) Bio-efficacy
   (iii) Toxicity
   (iv) Registration status in other countries.

Guidelines/Data Requirements For Provisional Registration Antagonistic Bacteria/Fungi and Entomogenous Fungi, Under Section 9 (3b) of the Insecticides Act, 1968

Mandatory Requirements

1. Submission of sample for pre-registration verification
2. Undertaking to be submitted

1. Standards of Formulations :

(i) CFU on selective medium minimum of 2×10^6 CFU per ml or gm for Trichoderma, Gliocladium spp. For Beauveria and other Entomogenous fungi CFU count should be minimum 1×10^8ml or gm.

(ii) Pathogenic contaminants such as Salmonella, Shigella or Vebrio should not be present. Other microbial contaminants not to exceed 1×10^4 count ml/gm.

(iii) Maximum moisture content should not be more than 8% for dry formulations of fungi and 12% for bacteria.

(iv) Stability or CFU bacteria counts at 30 deg. C and 65% for RH.

2. Registration Requirements

A. Chemistry

(i) Systematic name (Genus and species)

(ii) Common name, if any

(iii) Natural occurrence of the organism and morphological description

(iv) Composition and specification

(v) Manufacturing process: The microbial cultures are multiplied by liquid solid fermentation. Information pertaining to use of mycelial mats with spores separated, must be provided.

(vi) Test method

  (a) Dual culture for antagonistic fungi/bacteria

  (b) Pathogenicity test on insect in case of entomopathogenic fungus.

(vii) Qualitative analysis

  (a) CFU on selective medium

  (b) Absence of gram negative bacterial contaminants (*Salmonella, Shigella and Vibrio)*

  (c) Maximum moisture content should not be more than 8% for dry formulation of fungi and 12% for bacteria

  (d) Shelf-life claims – The moisture content in the biopesticides should not be exceeding 12% in the case of talc-based formulations of Trichoderma, Pseudomonas, etc. The shelf-life may be fixed at four months at room temperature not exceeding 32^0C. Beyond four months, it is not desirable to use the product as the concentration of the viable CFU of the organisms is drastically reduced. However, in case of formulations based on organic substances, the moisture content may go up to 30%.

B. Bioefficacy

(i) Laboratory test : The product should be tested at one laboratory under

ICAR/SAU/CSIR/ICMR system.

(ii) Field test : Results of some trials conducted under Indian conditions.

(iii) Data on non-target organisms ; one season/one year data on effect of product against natural parasites / predators in case of entomogenous fungi.

C. Toxicity

(a) For formulated product to be directly manufactured

(Mammalian Toxicity Testing of Formulations)

(i) SINGLE DOSE ORAL (Rat and mouse) Toxicity / Infectivity / Pathogenicity)

(ii) SINGLE DOSE PULMONARY Toxicity / Infectivity / Pathogenicity)

(iii) SINGLE DOSE DERMAL Infectivity

(iv) SINGLE DOSE INTRAPERITONEAL (infectivity)

(v) PRIMARY SKIN & EYE IRRITATION

Note : Considering the fact that many small entrepreneurs are engaged in the business of cultivation of NPV/GV, the following simplification has been suggested.

1. If the same primary culture is used for making the formulation by different entrepreneurs then the information submitted once on primary culture will be sufficient. All entrepreneurs need not submit the data individually.
2. If same primary culture, same method and same adjuvants, stabilizers etc., are used for making the formulation, then data once submitted will be sufficient for subsequent registrants.

(b) For primary culture :

(i) Single dose oral (rat and mouse)

(ii) Single dose pulmonary

(iii) Single dose dermal

(iv) Single dose Intraperitonial

(c) For formulation : In addition to data on primary culture, data on following may be required.

(i) Data on primary culture with technical

(ii) Single dose oral (rat and mouse)

(iii) Single dose pulmonary

(iv) Primary skin irritation

(v) Primary eye irritation

(d) Packaging and Labelling:

The packaging system is important to maintain desired moisture content. Therefore, the manufacturers should clearly specify the packaging material and the containers to be used.

(a) Packaging :

(i) Classification – solid, liquid or other types product.

(ii) Unit pack size- in metric system.

(iii) Specification: Details of Primary, Secondary & Transport pack.

(iv) Compatability of primary pack with the product (Glass bottles are not recommended).

(b) Labels and leaflets : As per Insecticides Rules, 1971/as per existing norms indicating the common name, composition, antidote/storage statements etc., handling, storage and transportation.

(c) No glass or fragile container shall be proposed.

Undertaking By Importers / Manufacturers of Microbial Pesticides

I, ______________________ aged ________________ years, S/o ____

Resident of __________________________ and ____________________ of M/s ________________________________ Registrated Office at ________________________________ do hereby undertake as follows.

(a) That the product ________________________________ based on ______________________, Strain __________________ manufactured by M/s ____________________ does not contain any genetically modified organism (GMO).

(b) That I / we shall abide by the provisions contained in the International Plant Protection Convention with regard to the import of this product.

(c) That I/we shall abide by the provisions in context of International Standards for PhytoSanitary Measures – Code of Conduct for the import and release of exotic biological Control agents of the International Plant Protection Convention (IPPC), FAO, Rome.

(d) That I/We shall provide the samples of our __________________ products as and when desired by the competent Authorities of Government of India for verification.

(e) That I / We further undertake that in the event of the above product having proved otherwise by any competent Authority and resulting in environmental damage, I/We shall inform the Central Insecticides

Board & Registration Committee, the Manufacturing Licensing Authority, Pollution control Authorities at appropriate District / State / National level and shall comply with the directions / decisions from them.

(f) That my/our above undertaking is true, and no portion is false and I have concealed nothing relevant to the above matter.

Date ______________________ Signature :

Place ______________________ Name ______________

Designation __________

Seal of the Company

Guidelines/ Data Requirements For Regular Registration Of Antagonistic Bacteria / Fungi And Entomogenous Fungi, Under Section 9 (3) Of The Insecticides Act, 1968

Mandatory Requirements :

1. Submission of sample for pre-registration verification
2. Undertaking to be submitted as per Annexure-III

1. Standards of Formulations

(i) CFU on selective medium minimum of 2×10^6 CFU per ml or gm for Trichoderma, Gliocladium spp. For Beauveria and other Entomogenous fungi CFU count should be minimum 1×10^8ml or gm.

(ii) Pathogenic contaminants such as Salmonella, Shigella or Vebrio should not be present. Other microbial contaminants not to exceed 1×10^4 count ml/gm.

(iii) Maximum moisture content should not be more than 8% for dry formulations of fungi and 12% for bacteria.

(iv) Stability or CFU bacteria counts at 30 deg. C and 65% for RH.

2. Registration Requirements

A. CHEMISTRY

(i) Systematic name (genus and species),

(ii) Common name, if any,

(iii) Natural occurrence of the organism and morphological description,

(iv) Composition and specification,

(v) Manufacturing process: The microbial cultures are multiplied by liquid solid fermentation. Information pertaining to use of ensure mycelial mats with spores separated, must be provided.

(vi) Test method

(a) Dual culture for antagonistic fungi/bacteria

(b) Pathogenicity test on insect in case of entomopathogenic fungus.

(vii) Qualitative analysis.

(a) CFU on selective medium

(b) Absence of gram negative bacterial contaminants (*Salmonella, Shigella and Vibrio)*

(c) Maximum moisture content should not be more than 8% for dry formulation of fungi and 12% for bacteria

(d) Shelf-life claims – The moisture content in the biopesticides should not be exceeding 12% in the case of talc-based formulations of Trichoderma, pseudomonas, etc. The shelf-life may be fixed at four months at room temperature not exceeding 32^0C. Beyond four months, it is not desirable to use the product as the concentration of the viable CFU of the organisms is drastically reduced. However, in case of formulations based on organic substances, the moisture content may go up to 30%.

B. Bioefficacy

(i) Laboratory test : The product should be tested at one laboratory under

ICAR/SAU/CSIR/ICMR system.

(ii) Field test : Results of some trials conducted under Indian conditions.

(iii) Data on non-target organisms ; one season/one year data on effect of product against natural parasites / predators in case of entomogenous fungi.

C. Toxicity

(a) For formulated product to be directly manufactured

(Mammalian Toxicity Testing of Formulations)

(i) SINGLE DOSE ORAL (Rat and mouse)
Toxicity / Infectivity / Pathogenicity)

(ii) SINGLE DOSE PULMONARY
Toxicity / Infectivity / Pathogenicity)

(iii) SINGLE DOSE DERMAL
Infectivity

(iv) SINGLE DOSE INTRAPERITONEAL (infectivity)

(v) PRIMARY SKIN & EYE IRRITATION

Note : Considering the fact that many small entrepreneurs are engaged in the business of cultivation of NPV/GV, the following simplification has been suggested:

1. If the same primary culture is used for making the formulation by different entrepreneurs then the information submitted once on primary culture will be sufficient. All entrepreneurs need not submit the data individually.
2. If same primary culture, same method and same adjuvants, stabilizers etc., are used for making the formulation, then data once submitted will be sufficient for subsequent registrants.

(b) For primary culture
   (i) Single dose oral (rat and mouse)
   (ii) Single dose pulmonary
   (iii) Single dose dermal
   (iv) Single dose Intraperitonial

(c) For formulation : In addition to data on primary culture, data on following is required.
   (i) Data on primary culture with technical
   (ii) Single dose oral (rat and mouse)
   (iii) Single dose pulmonary
   (iv) Primary skin irritation
   (v) Primary eye irritation

(d) Packaging and Labelling

The packaging system is important to maintain desired moisture content. Therefore, the manufacturers should clearly specify the packaging material and the containers to be used.

(a) Packaging
   (i) Classification – solid, liquid or other types product.
   (ii) Unit pack size in metric system.

(iii) Specification: Details of Primary, Secondary & Transport pack.

(iv) Compatability of primary pack with the product (Glass bottles are not recommended).

(a) Labels and leaflets : As per Insecticides Rules, 1971/as per existing norms indicating the common name, composition, antidote/storage statements etc. handling, storage and transportation).

(b) No glass or fragile container shall be proposed.

Undertaking By Importers / Manufacturers of Microbial Pesticides

I, ____________________________ aged ____________ years, S/o ______________________________________

Resident of _________________________ and ______________________ of M/s __________________________________ Registrated Office at __________________________ do hereby undertake as follows.

(a) That the product ______________________________________ based on ___________________, Strain ______________________ manufactured by M/s _____________________ does not contain any genetically modified organism (GMO).

(b) That I / we shall abide by the provisions contained in the International Plant Protection Convention with regard to the import of this product.

(c) That I/we shall abide by the provisions in context of International Standards for Phytosanitary Measures – Code of Conduct for the import and release of exotic biological Control agents of the International Plant Protection Convention (IPPC), FAO, Rome.

(d) That I/We shall provide the samples of our ___________________________ products as and when desired by the Competent Authorities of Government of India for verification.

(e) That I / We further undertake that in the event of the above product having proved otherwise by any competent Authority and resulting in environmental damage, I/We shall inform the Central Insecticides Board & Registration Committee, the Manufacturing Licensing Authority, Pollution Control Authorities at appropriate District / State / National level and shall comply with the directions / decisions from them.

(f) That my/our above undertaking is true, and no portion is false and I have concealed nothing relevant to the above matter.

Date ______________________ Signature :

Place ______________________ Name ______________

Designation __________

Seal of the Company

Guidelines/Data Requirements For Regular Registration Of Antagonistic Bacteria/Fungi And Entomogenous Fungi, Under Section 9 (3) Of The Insecticides Act, 1968

1. Standards of Formulations

(i) CFU ON SELECTIVE MEDIUM MINIMUM OF 2×10^6 CFU per ml or gm for Trichoderma, Gliocladium spp. For Beaiveria and other Entomogenous fungi CFU count should be minimum 1×10^8/ml or gm.

(ii) Pathogenic contaminants such as Salmonella, Shigella or Vebrio should not be present. Other microbial contaminants not to exceed 1×10^4 count ml/gm.

(iii) Maximum moisture content should not be more than 8% for dry formulation of fungi and 12% for bacteria.

(iv) Stability or CFU bacteria counts at 30 deg. C and 65% for RH.

2. Registration Requirements

A. Chemistry

(i) Systematic name (Genus and species)

(ii) Common name, if any

(iii) Natural occurrence of the organism morphological description.

(iv) Composition and specification

(v) Manufacturing process: The microbial cultures are multiplied by liquid solid fermentation. Information pertaining to user of entire mycelial mats with spore separated, must be provided.

(vi) Test method:

1. Dual culture for antagonistic fungi/bacteria
2. Pathogenicity test on insect in case of entomopathogenic fungus.

(vii) Qualitative analysis:

(a) CFU on selective medium

(b) Absence of gram negative bacterial contaminants (Salmonella, Shigella and Vibrio)

(c) Moisture content should not be more than 8% for dry formulation of fungi and 12% for bacteria

(d) Shelf-life claims – The moisture content in the biopesticides should not be exceeding 12% in the case of talc-based formulations of Trichoderma, Pseudomonas, etc. The shelf-life may be fixed at four months at room temperature not exceeding 32 °C. Beyond four months, it is not desirable to use the product as the concentration of the viable CFU of the organisms is drastically reduced. However, in case of formulations based on organic substances, the moisture content may go up to 30%.

B. Bioefficacy

(i) Field tests : Result of two trails in two agro climatic conditions by ICAR/SAU/CSIR/ICMR Institutes. Data should be certified either by the University authority or Head of the Institute concerned.

(ii) Data on non-target organisms: one season/one year data on effect of product against natural parasites/predators in case of entomogenous fungi.

C. Toxicity

(a) For formulated product to be directly manufactured

(Mammalian Toxicity Testing of Formulations)

(i) SINGLE DOSE ORAL (Rat and mouse)
Toxicity / Infectivity/Pathogenicity

(ii) SINGLE DOSE PULMONARY
Toxicity/Infectivity/Pathogenicity
(Intratracheal preferred)

(iii) SINGLE DOSE DEERMAL
Infectivity

(iv) SINGLE DOSE INTRAPERITONEAL (Infectivity)
PRIMARY SKIN & EYE IRRITATION

(v) PRIMARY EYE IRRITATION TEST

(vi) HUMAN SAFETY RECORDS.

(vii) (Effect/Lack of effects)

Note : Considering the fact that many small entrepreneurs are engaged in the business of cultivation of NPV/GV, the following simplification has been suggested.

1. If the same primary culture is used for making the formulations by different entrepreneurs, then the information submitted once on primary culture will be sufficient. All entrepreneurs need not submit data.
2. If the same primary culture, same method and same adjuvants, stabilizers etc., are used for making formulations then data once submitted will be sufficient for subsequent registrants.

(b) For Primary culture

(i) Single dose oral (rat and mouse)

(ii) Single dose pulmonary

(iii) Single dose dermal

(iv) Single dose intraperitonial

(c) For formulation

(i) Data on primary culture with technical

(ii) Single dose oral (rat and mouse)

(iii) Single dose pulmonary

(iv) Primary skin irritation

(v) Primary eye irritation

(vi) Human safety Records

(d) ***Environmental Safety Testing : Core Information Requirements (for formulation only)***

(i) Non-target vertebrates, a. Birds (two species), b. Fresh water fish

(ii) Non-target Invertebrates, c. Terrestrial Invertebrates, d. Soil Invertebrates:

1. Information on infection and pathogenicity in mammals will be available from mammalian safety testing.
2. Information on infection and pathogenicity, suggested test; single-dose, oral test, Suggested test species; bobwhite, quail, Japanese quail, mallard duck/pigeon and chicken.

3. Information on infection and pathogenicity, Suggested test species; rain bow trout, blue gill sunfish, Tilapia mossambica.
4. Information on mortality effects. It is recommended that information be obtained for honey bee, silk worm.
5. Information on mortality effects. It is recommended that test species include an earthworm (Lumbricus terrestris) or other appropriate macroinvertebrates of ecological significance.

(e) ***Processing, Packaging and Labelling***

The packaging system is important to maintain the desired moisture content. Therefore, the manufacturers should clearly specify the packaging material and the containers to be used.

(a) Packaging

Classification-solid, liquid or other types product.

Unit pack size-in metric system.

Specification ; Details of Primary, Secondary & Transport pack.

Compatibility of primary pack with the product (Glass bottles are not recommended).

(b) Labels and leaflets: As per Insecticides Rules, 1971/as per existing norms indicating the common name, composition, antidote/storage statements etc.

(The packaging material should also be ensured free from contamination from handling, storage and transporatation.)

(c) No glass or fragile container shall be proposed.

Indian Standards Antagonistic Bacteria / Fungi And Entomogenous Fungi Specifications

1. Form and appearance
2. Moisture contents
3. pH
4. Composition
5. CFU counts: Trichoderma -2×10^6 CFU/ml or gm Entomopathogenic fungi-1 $\times 10^8$ CFU/ml or gm
6. Contaminations: Salmonella, Shigella or Vibro : absent. Other microbial Contaminants should not be exceed 1 $\times 10^4$ ml or gm.

7. Method of analysis:
   (a) CFU counts by serial dilution and examination under phase contrast microscope.
   (b) Plating for contaminants on specific media.
   (c) Antagonistic capacity on target microscope by bioassay on plates.
   (d) Entomopathogenic capability on target insects by bioassays.
8. An undertaking should be submitted that the strain is indigenous, naturally occurring, not exotic and not genetically modified.

Annexures

Table 1 Sections / groups within the genus *Trichoderma* (Bisset, 1991 a, b; Gams and Meyer, 1998; Samuels et al., 1999; Dodd et al.,2003).

Section (group)	Characteristics	Species aggregate
Trichoderma (T.atroviride – T.koningii complex or T.atroviride – T.viride complex)	The species in this section have narrow and floxuos conidiophores with branches and phialides uncrowded, frequently paired and seldom in verticils of more than three.	*T.viride, T.koningii, T. aureoviride, T.atroviride T.asperellum*
Longibrachiatum (*T.longibrachiatum* group)	In this section the species have conidiophores which are sparingly and irregularly branched with the phialides also irregularly disporsed and not usually in whorls or vertical. Species may also produce distinctive yellow greenish pigments in the reverse of cultures.	*T.longibrachiatum, T.pseudokoningii, Trichoderma anamorphs of H.schweinitzil*
Saturnisporum	The species have a *T. viride* like branching pattern with branches and phialiades uncrowded and frequently paired but with inflated phialides and compact conidiogenous pustules as in section Pachybasium. They are further differentiated by the bullate or wing like conidial ornamentation.	*T.saturnisporum, t.ghanensi*
Pachybasium (*T.harzianum – T.hamatum* complex)	It includes species with highly ramified, broad conidiophores usually arranged in compact postules or fascicles and with branches and phialides broad or inflated, relatively short and disposed in crowded verticils. Some species are further characterized by the production of sterile conidiophore extensions and many isolates produce compact conidiogenous pustules with adjacent conidiophore anastomosing.	*T.hamatum, T.inhamatum, T.polysporum, T.piluliferum, T.harzianum, T.filanitosa, fasciculatum, T.flavofuscum, T.crassum, T.virens, T.croceum, T.semorbis, T.minutisporum, T.tomentosum, T.fertile, T.spirale, T.longipilis, T.oblongisporum, T.stringosum, T.strictipillis, T.pubescens, T.aggressivum *, T.aggressivum f.europaeum*.*

Table 1 Contd...

Section (group)	Characteristics	Species aggregate
Hypocreanum	It accommodates Hypocrea anamorphs, mostly from section Homalocrea and occasional isolates from soil or wood that are characterized by effuse, usually sparse conidiation, sparingly branched conidiophore and cylindrical to subulate phialides frequently borne in Verticillium – like divergent verticals	*Hypocreanam*

Table 2 Suppression / management of plant pathogens / disease by *Trichoderma* spp. (Singh et al., 2003)

Crop	**DfDisease**	**Pathogen (s)**	**Trichoderma spp.**
Fruit crops			
Apple	White root rot	*Dematophora necatrix*	*Trichoderma viride, T.harzianum, T.virens*
Citrus (Mandarin)	Root rot	*Phytophthora nicotinae pv.parasitica, p.colocasiae*	*T.harzianum, T.virens, T.viride.*
Guava	Anthracnose	*Collectotrichumk gloeosporioides, Pestalotia psidii*	*T.harzianum*
	Fruit rot	*Lasiodiploidia theobromae, C.gloeosporioides, Pestalotiopsis versicolor, Phomopsis psidi, Rhizopus arhizus*	*Trichoderma spp.*
Mango	Fruit rot	*Lasiodiplodia theobromae, Rhizopus arrhinus*	*Trichoderma spp.*
Melon	Wilt	*F.oxysporum*	*T.viride*
Muskmelon	Wilt	*F.oxysporum*	*T.harzianum*
Orange	Blue mould	*Penicillium italicum*	*T.harzianum*
Strawberry	Grey mould	*B.cineria*	*T.harzianum*
Water melon	Wilt	*F.oxysporum f.sp. solani*	*T.viride*

Table 2 Contd...

Crop	DfDisease	Pathogen (s)	Trichoderma spp.
Bean	*Seedling rot*	*Pythium sp., s.sclrotiorum, B:P cineria, R.solani*	*T.koningii*
Cabbage	*Alternaria blight*	*Alternaria brasicicola*	*T.virens, T.longibrachiatum*
Cabbage	*Damping off*	*Rhizoctonia solani*	*T.harzianum, T.viride, T.koningii*
Cauliflower	*Damping off*	*R.solani, Pythium aphanidermatum*	*T.harzianum*
Chilli	*Root rot*	*S.rolfsii*	*T.harzianum*
	Fruit rot		
	die back	*Colletotrichum capsici*	*T.viride, T.harzianum, T.koningii, T.pseudokoningii, T.hanatum, T.longibrachiatum, T.pileatus,*
Egg plant	*Wilt, damping off*	*F.solani, Pythium aphanidernatun*	*T.harzianum, T.viride, T.koningii*
	Collar rot	*Sclerotinia sclerotiorum*	*T.virens, T.viride*
Fenugreek	*Root rot*	*R.solani*	*Trichoderma spp.*
French bean	*Root rot*	*R.solani*	*T.virens, T.harzianum, T.hamatum, T.viride*
Pea	*Seed and collar rot*	*Pythium sp., R.solani*	*T.harzianum, T.hamatum*
	White rot	*Sclerotinia sclerotiorum*	*T.harzianum, T.viride, T.virens*
Potato	*Black-scurf*	*R.solani*	*T.virens, T.viride*

Table 2 Contd...

Crop	DfDisease	Pathogen (s)	Trichoderma spp.
Radish	*Seedling-rot, Damping – off*	*Pythium sp., R.solani*	*T.harzianum, T.hamatum*
	Seed rot	*R.solani*	*T.harzianum*
Tomato	*Damping off*	*Pythium indicum*	*T.viride, T.harzianum*
	Grey mould	*B.cineria*	*T.harzianum*
	Damping off	*Pythium aphanidermatum, R.solani T.pseodokoningii*	*T.viride, T.koningii,*
	Seedling disease and wilt	*F.oxysporum f.sp. lycopersici*	*Trichoderma spp.*
	Root knot	*Meloidogyne incognita, M.javanica*	*T.harzianum*
Oilseed crops			
Groundnut	*Wilt complex seed rot*	*Sclerotium rolfsii, F.solani, F.oxysporum S.rolfsii*	*T.harzianum, T.virens, T.viride*
	Root and Stem rot	*R.solani*	*T.virens, T.longibrachiatum*
	Leaf rust	*Puccinia arachidis*	*T.harzianum*
	Collar/Root/Crown/ Stem/Pod rot	*A.flavus, S.rolfsii, A.niger*	*T.harzianu, T.viride, T.viride*
Mustard	*Damping off*	*Pythium aphanidermatum,*	*T.harzianum, T.viride*
Safflower	*Root rot*	*M.phaseolina*	*T.viride*
Sesamum	*Blight*	*Phytophthoraa sp.*	*T.harzianum, T.viride*
	Root rot	*M.phaseolina*	*T.harzianum, T.viride, T.longibrachiatum, T.virens*
Sunflower	*Blight*	*Alternaria helianthi*	*T.virens*
	Root/collar rot	*Sclerotium rolfsii, R.solani, Sclrotinia sclerotiorum*	*T.harzianum, T.hatum*

Table 2 Contd...

Crop	DfDisease	Pathogen (s)	Trichoderma spp.
Fibre crop			
Cotton	*Root rot*	*M.phasiolina*	*T.harzianum, T.virens*
	Wilt	*F.oxysporum f.sp. vasinfectum, S.rolfsii R.solani*	*T.harzianum*
Plantation crops			
Arecanut palm	*Fruit rot*	*Phytophthora spp.*	*Trichoderma spp.*
Coffee	*Collar rot*	*R.solani*	*T.harzianum*
Mulberry	*Leaf spot*	*Cercospora moricola*	*T.harzianum, T.viride*
	Stem canker & die-back	*Botryodiploidia spp.*	*T.harzianum, T.virens, T.pseudokoningii*
	Cutting rot	*F.solani*	*T.harzianum, T.virens, T.pseudokoningii*
	Collar rot	*Phona sorghina*	*T.harzianum, T.virens, T.pseudokoningii*
Rubber	*Brown root*	*Phellinus noxius*	*T.harzianum, T.virens, T.hamatum*
Sugar crops			
Sugar beet	*Damping off*	*Pythium aphanidermatum*	*T.harzianum*
	Root rot /wilt	*S.rolfsii*	*T.harzianum*
Sugarcane	*Root rot*	*Pythium graminicola*	*T.viride*
	Red rot	*Colletotrichum falcatum*	*T.harzianum, T.viride, plfluorescens*
	Seedling rot	*Pythium graminicola*	*T.viride*
	Wilt	*F.moniliformae*	*T.harzianum, T.viride, T.longibrachiatum,*

Table 2 Contd...

Crop	**DfDisease**	**Pathogen (s)**	**Trichoderma spp.**
Condiments, Spices, Narcotics, Stimulants etc.			
Cardamom	*Capsule rot*	*Phytophthora meadii*	*T.harzianum, T.viride*
	Damping off	*Fusarium moniliforme, Pythium vexans, Rhizoctonia sp., Pythium aphanidernmatum*	*T.harzianum, T.viride*
Cariander	*Wilt*	*F.oxysporum f.sp. corianderii*	*T.viride, T.harzianum*
Gringer	*Rhizome rot*	*F.oxysporum f.sp., zingiberi*	*T.harzianum, G.virens*
	Rhizome rot	*Pythium pleroticum, F.equiseti, Pythium myriotylum, F.solani*	*T.viride*
	Yellows	*F.oxysporum f.sp. zingiberi*	*T.harzianum, T.virens, T.hamatun*
	Rhizome rot	*S.rolsfsii, Fusarium solani, Pythium sp.*	*T.harzianum, Trichoderma sp. T.virens*
Pepper	*Collar rot*	*Phytophthora capsici*	*T.virens, Trichoderma spp.*
Poppy (Opium)	*Downy mildew*	*Peronospora arborescens*	
Tobacco	*Damping off*	*Pythium aphanidermatum*	*T.harzianum*
Cereals			
Barley	*Foot and root rot*	*Sclerotium rolfsii, Fusarium, Curvularia, Pythium, Aspergillus, Penicillium*	*T.viride, T.pseudokoningii*
Maize	*Charcoal rot, Banded blight*	*M.phaseolina, R.solani*	*Trichoderma spp.*
Rice	*Shealth blight*	*R.solani*	*T.virens, T.harzianum, T.viride*
	Kernel smut	*Tilletia barclayana*	*T.harzianum, T.viride, T,virens, T.deliquescens*

Table 2 Contd...

Crop	DfDisease	Pathogen (s)	Trichoderma spp.
	Bunt	*Neovossia indica*	*T.harzianum, T.viride, T.virens, T.deliquescens*
	Brown spot	*Drechslera oryzae*	*T.viride*
	Sheath blight	*R.solani*	*T.viride, T.harzianum, T.longibrachiatum, T.aureoviride*
Sorghum	*Seed pathogens*	*A.flavus, A.niger, Alternaria alternata, Cladosporiu7m olysporum, Collectrotricum dematium, Cuvularia lunata, Exserohilum halodes, F.moniliforme, Phomasogina*	*T.harzianum, T.viride*
Wheat	*Root rot*	*S.rolfisii, F.oxysporum*	*T.harzianum*
	Takeall	*Gaeumanomyces graminis var. tritici*	*T.harzianum*
	Karnal bunt	*Neovossia indica*	*T.viride, T.harzianum, T.koningii T.virens, T.deliquescens.*
	Loose smut	*Ustilago segatun tritici*	*T.viride, T.koningii, T.harzianum, T.lignorum*
	Spot blotch	*Drechslera sorokiniana*	*T.reesei, T.psudokoningii, T.viride.*
Flowering plants			
Gladiolus	*Yellows and Corm rot*	*F.oxysporum gladioli*	*T.virens*
Jasmine	*Root rot*	*M.phaseolina*	*T.harzianum, T.viride*
Rose	*Grey mould*	*Botrytis cineria*	*T.harzianum, T.viride*

Table 2 Contd...

Crop	**DfDisease**	**Pathogen (s)**	**Trichoderma spp.**
Pulses			
Blackgram	*Dry root rot*	*Macrophomina phaseolina*	*Trichoderma harzianum*
	Damping off	*Sclerotium rolfsii*	*T.viride*
Chickpea	*Wilt, Seed rot. Root rot*	*F.oxysporum f.sp. ciceris, R.bataticola, Pythium sp.*	*T.harzianum*
	Wilt-complex	*F.oxysporum f.sp. ciceris, R.solani, S.rolsfsii.*	*T.virens, T.harzianum, T.viride*
	Gray mould	*Botrytis cineria*	*Trichoderma spp.*
	Collar rot	*S.rolfsii, R.solani*	*T.harzianum, T.viride*
	Root rot	*Fusarium solani and M.phaseolina*	*T.virens, T.harzianum, T.viride*
	Stem rot	*Sclerotinia sclrotiorum*	*T.harzianum*
	Wilt	*F.oxysporum f.sp. ciceris*	*T.virens, T.harzianum, T.viride*
Cowpea	*Wilt*	*Fusarium oxysporum f.sp. tracheiphilum*	*T.harzianum and T.viride*
	Charcoal rot and wilt	*M.phaseolina, Fusarium oxysporum f.sp. tracheiphilum*	*T.viride, T.harzianum, T.koningii, T.pseudokoningii*
Horse gram	*Damping-off*	*M.phaseolina*	*T.viride*
Lentil	*Wilt complex: collar rot*	*R.solani, F.oxysporum, S.rolfsii*	*T.virens*
Moth bean	*Blight*		*T.harzianum, T.viride*
Mung bean	*Root rot*	*M.phaseolina*	*T.harzianum, T.viride*
	Bacterial leaf spot	*Xanthomonoas compestris pv. Vgnae radiatae*	*T.harzianum, T.viride*

Table 3 Commercial formulations of biocontrol agents available in India.

Product	Bio-agents (s)	Developing agency
Antagon-TV	T.viride	Green Tech Agro Products, Combatore
Biocon	T.viride	Tocklai Experimental Station, Tea Research Association, Jorhat, ASSAM
Bioderma	T.viride + T.harizianum	Biotech International Limited, New Delhi
Bioguard	T.viride	Krishi Rasayan Export Pvt. Ltd., Solan (H.P.)
Bioshield	Pseudomonas fluorescens	Anu Biotech International Ltd., Faridabad
Biotok	Bacillus subtilis	Tocklai Experimental Station, Tea Research Association Jorhat, ASSAM
Ecoderma	T.viride + T.harzianum	Margo Biocontrol Pvt. Ltd., Bangalore
Ecofit	Trichoderma viride	Hoechst and Schering AgrEvo Ltd., Mumbai
Fungintil	T.viride	Crop Health Bioproduct Research Centre, Gaziabad.
Kalisena SD Kalisena SL	Aspergillus niger AN-27	Cadila Pharmaceuticals Limited, Ahmedabad

List of commercial producers of Biocontrol agents in India

S.No.	Name and Address of the Company	Bio-agents produced
1.	Agriland Biotech Limited Office NO.36, Prince Ind. Estate, MotaMotipura, Baroda- 391 520, Gujarat.	*Helicoverpa armigera NPV; Beuvaria bassiana, Pacilomyces lilacinus, Trichoderma viride, Spodoptera litura NPV.*
2.	Margo Biocotnrols Pvt. Ltd. 344/8, IV Main, Sadashivanagar, Bangalore- 560 080.	*Ha NPV, Verticillium lecanii, T.viride*
3.	Biotech International Ltd. VIPPS Centre, Local Shopping Centre Block-EFGH, Masjid Moth,Greater Kailash-II, New Delhi-110048.	*Ha NPV, Trichoderma chilonis, T.japonicum, Pseudomonas fluorescens, T.viride, Sl NPV, Beauveria bassiana*
4.	Varsha Bioscience & Technology 17-1-382/SN/1/1, Balaji Nagar, J.S.Reddy Sports Club Road, Champapet, Saidabad Hyderabad-5000 050, A.P.	*Ha NPV, P.fluorescens, T.viride*
5.	Basarass Biocon (India) Pvt. Ltd. 24/7, Dr.radhakrishnan Salai, Palaniappa Nagar, Valasaravakkam, Chennai – 600 087, Tamilnadu	*Ha NPV, Sl NPV, B.bassiana, M.anisopliae, T.chilonis, Chrysopids, Cryptolaemus T.harzianum, T.viride, P.fluorescens, Metarhizium anisopliae*
6.	Bio-control Research Laboratories (BCRL), P.O.Box No. 6426, Yelahanka Post office, 36/2, Sriramanahalli, Near Rajanakunte Bangalore – 560 064, Karnataka.	*Ha NPV, T.chilonis, T.haponicum, Chrysopids, Cryptolaemus, Goniozus, P.fluorescens Trichoderma harzianum, T.viride, Sl NPV*
7.	Bio-pest Mangemetn Pvt.Ltd. 12th Km. Opp. Jakkur Aerodrome, Jakkur Post, Bangalore-560 064, Karanataka	*Ha NPV, T.chilonis, Chrysopids, T.viride, Sl NPV.*
8.	Nominagri Bio-Pvt. Ltd S.No. 79/2/2, Ambervet, TW Mulshi, Dist. Puna, Maharashtr	*Ha NPV ; B.basiana, T.chilonis, Chrysopids, Cryptolaemus, V.lecanii, P.lilacinus, P.fluorescens, T.viride*

Table 4 Contd...

S.No.	Name and Address of the Company	Bio-agents produced
9.	Sribiotech 112, Sri Arcade, BATA Complex Erragadda, Hyderabad-18, AP	*Ha NPV, B.bassiana, M.anisopliae, V.lecanii, P.fluorescens, T.viride*
10.	KN Bio Sciences India Pvt. Ltd. Plot No.55, B.H.E.L HIG Phase II Ushodaya Enclave, Miyapur Hyderabad-49, AP	*b.bassina, M.anisopliae, v.lecanii, P.lilacinus, P.fluorescens, T.viride, Sl NPV*
11	R.B. Herbal Agro Trimurti Kacheri Road, Satana Dt. Nashik, Maharashtra, Pin-423301.	*B.bassiana, M.anisopliae, V.lecanii, P.fluorescens, T.viride*
12.	Sovereign Plot NO.6, Block No.38, Phase IV Indira Auto Nagar, Guntur, AP	*B.bassiana V.lecanii, P.fluorescens, T.viride*
13.	Plantrich Industrial Estate, Manarcadu P.O. Kottayam-686 019, Kerala	*B.bassiana, M.anisopliae, V.lecanii, P.lilacinus, P.fluorescens, T.harzianum, T.viride*
14.	Greenstar Agro Corporation 53, Subramanyam Street, K.N.K. Road Back Side, Erode-638 003, TN	*P.fjuorescens, T.viride*
15.	Bacto Agro Culture Care (p) Ltd. Bharatha Nagar, Kolappally pandalur Taluk, The Nilgiris, TN	*P.fluorescens, T.harzianum*
16.	State Biocontrol laboratory Mannuthy P.O., Thrissur – 680 651, Kerala	*T.chilonis, P.fluorescens*
17.	Soman Bio Fertilizers 594, Sadashiv Peth, 101, Express Topwer Laxmi Road, Pune-30 Maharashtra	*V.lecanii, T.viride*

Table 4 Contd...

S.No.	Name and Address of the Company	Bio-agents produced
18.	Biopesticide production Unit Krishi Vigyan Kendra Regional Agril.Resrarch Station Kerala Agril.University Kottayam – 686 566, Kerala.	*P.fluorescens, T.viride*
19.	Rajendra Foundation for Agril. Research & Rural Devep. Chittar, Bhavani (TK) Erode (Dt.) – 638 311, TN	*T.chilonis, T.japonicum* *T.brassiliense, T.achaeae,* *Chrysopids, Cryptolaemus,* *Goniozus*
20.	Shakthi Biotech Pvt. Ltd. Ampapuram (opt to TATA Tower) Near Telaprolu Krishna (Dt)-521 109, AP	*T.viride*
21.	Sun Agro Biosystems Pvt. Ltd. 3/340 Main Road, madanandapuram Porur, Chennai – 600 116, TN	*B.bassiana, M.anisopliae,* *T.chilonis, P.fluorescens,* *T.viride*
22.	Greentech Agro Services Pvt. Ltd. 47-D, PARSNS Resisdency Rajaji Road, Ram Nagar Coimbatore-641 009, TN	*P.lilacinus, P.fluorescens,* *T.harzianum*
23.	Varanashi Agro Sustainable Technology Center Adyanedka – 574024, Karnataka	*T.harzianum*

Bibliography

Agarwal SC, Khare MN and Agarwal PS. 1977. Biological control of *Sclerotium rolfsii* causing collar rot of lentil. *Indian Phytopathology*, **30**: 176-177.

Ahmad JS. and Baker R. 1987. competitive saprophytic ability and cellulolytic activity of rhizosphere – competent mutants of *Trichoderma harzianum*. *Phytopathology,* **77:** 358-362.

Ahmed El-Ghaouth, Joseph, L. Smilanick, Eldon Brown, G. Antonio Ippolito, Michael Wisniewski and Charles L. Wilson. 2000. Application of *Candida saitoana* and glycochitosan for the control of post harvest diseases of apple and citrus fruits under semi-commercial conditions. *Plant Disease,* **84**(3) : 243–248.

Alstrom, S. 1991. Induction of disease resistance in common bean susceptible to halo blight bacterial pathogen after seed bacterization with rhizosphere pseudomonads. *Journal of General and Applied Microbiology* **37**: 495-501.

Allen MF., Moore TS. Jr., and Christensen M. 1982. Phytohormone change in *Bouteloua gracilis* infected by vesicular – arbuscular mycorrhiza II: Altered levels of gibberellin- like substances and abscisic acid in the host plant. *Canadian Journal of Botany,* **60** : 468-471.

Andrews JH.(1992) Biological control in the phyllosphere. *A. Rev. Phytopathol.* **30,** 603-605.

Aneja KR. 1998. Biological suppression of aquatic weeds with fungal pathogens. Biological suppression of plant diseases, phytoparasitic nematodes and weeds. Singh S.P and Hussaini.S (eds.) Technical document p. 174-191.

Aneja KR. 1998. Biotechnology for the production and enhancement of mycoherbicide potential. In: From Ethnomycology to fungal biotechnology. Singh J and Aneja KR(eds.), Plenum Publishers, UK.

Aneja KR., Singh K. and Bhan R. 1988. A new leaf spot disease of water hyacinth. *Indian Phytopathology* **41** : 160.

Aneja KR. and Srinivas B. 1990. Leaf spot disease of water hyacinth, *Eichhornia crassipes* – a new disease record from India. *Tropical Pest Management* **36:** 405-406.

Asghari MA. and Mayee CD. 1991. Comparative efficacy of management practices on stem and pod rots of groundnut. *Indian Phytopathology* **44**: 328-332.

Atkinson D, Berta G. and Hooker, JE. 1994. Impact of mycorrhizal colonization on root longevity and the formation of growth regulators. In: Impact of arbuscular mycorrhizas on sustainable agriculture and natural ecosystems. Birkhauser, Basel. pp. 89-99.

Avis T J, Caron SJ, Boekhout T, Hamelin RC. and Belanger RR. 2001. Molecular and physiological analysis of the powdery mildew antagonist *Pseudozyma flocculosa* and related fungi. *Phytopathology* **91**(3) : 249–254.

Azcon Aguilar, C. and Barea, J. M. 1996. Arbuscular mycorrhizas and biological control of soil borne plant pathogens- an overview of the mechanisms involved, *Mycorrhiza*, **6:** 220-223.

Baby UI and Manibhushanrao K.1993. Control of rice sheath blight through the integration of fungal antagonists and organic amendments. *Tropical Agriculture* **70:** 240-244.

Backman PA, Brannen PM and Mahaffee WF. 1994. Plant response and disease control following seed inoculation with Bacillus subtilis. In: Improving Plant Productivity with Rhizosphere Bacteria. Ryder MH, Stephens PM and Bowen GD (eds.)CSIRO, Australia. Pp 3-8.

Bagyaraj DJ. 1984. Biological interactions with VA mycorrhizal fungi. In: VA mycorrhiza, CRC press, Boca Raton, pp. 131-153.

Bagyaraj DJ. 1989. Competitions among VAM fungi and their interactions with soil organisms. In: *Recent Advances in Microbial Ecology*, Japan Scientific Society Press, Japan, pp.237-241.

Bagyaraj DJ. 1989. Mycorrhizae. In: Tropical Rain Forest Ecosystems, Elsevier Science Publishers. Amsterdam, pp. 537-546.

Bagyaraj DJ. 1992. Vesicular arbuscular mycorrhiza- Application in agriculture. *Methods in Microbiology*, **24**: 359-374.

Bagyaraj DJ. 1996. Mycorrhizal symbiosis in tropical trees in relation to control of root pathogens. In: Impact of diseases and insect pests in tropical forests, Kerala Forest Research Institute, Peechi, India, pp.239-245.

Bagyaraj DJ. 2006. VA mycorrhizal fungi and their role in plant disease control. In: Current status of Biological Control of Plant Diseases using Antagonistic Organisms in India. B. Ramanujan and RJ Rabindra, eds. Technical Document, Project Directorate of Biological Control, Bangalore. Pp.125-134.

Bagyaraj DJ , Manjunth A and Reddy DDR.1979. Interaction of vesicular arbuscular mycorrhiza with root-knot nematodes in tomato. *Plant and Soil* **51**: 397-403.

Bagyaraj DJ., and Varma V. 1995. Interaction between arbuscular mycorrhizal fungi and plants: their importance in sustainable agriculture and in arid and semiarid tropics. In: *Advance in Microbial Ecology*, Academic Pres, London, pp. 119-142.

Baker, R. 1991. Diversity in Biological control. *Crop prot.* 10, 85-94

Baker, R. and Dunn, P.E. 1990. New directions in Biological control. Alternatives for suppressing Agricultural Pests and Diseases. Alan R. Liss. New York, 837pp.

Baker, K.F. 1987. Evolving concepts of biological control of plant pathogens.*A.Rev.Phytopathol.***25 :** 76-85.

Baker KF and Cook RJ.(1974) Biological control of plant pathogens. Fruman, San Francisco, California, 433pp.

Barea JM. and Jeffries P. 1995. Arbuscular mycorrhizas in sustainable soil plant systems. In: *Mycorrhiza- structure, function, molecular biology and biotechnology,* Springer, Heidelberg, pp. 521-529.

Beagle- Ristaino JE and Papavizas GC. 1985. Biological control of *Rhizoctonia* stem canker and black scurf of potato. *Phytopathology* **75:** 560-564.

Becker JO and Cook RJ. 1988. Role of siderophores in suppression of *Pythium* species and production in increased growth response of wheat by fluorescent pseudomonads. *Phytopathology* **78:** 778-782.

Benhamou N, Belanger RR and Pulitz T. 1996a. Ultrastructural and cytochemical aspects of the interaction between *Pseudomonas fluorescens* and Ri T-DNA transformed pea roots: host response to colonization by *Pythium ultimum* Trow. *Planta* **199**: 105-117.

Bhan VM, Kauraw LP and Archana Chile. 1998. Biological suppression of weeds with pathogens- present scenario. Biological suppression of plant diseases, phytoparasitic nematodes and weeds. Singh S.P and Hussaini.S (eds.) Technical document p. 164-173.

Blakeman JP and Fokkema NJ. 1982. Potential for biological control of plant diseases on the phylloplane. *Annual Review of Phytopathology* **20:** 167-192.

Brain B. Mispadden Gardener, Dmitri V.Mavrodi, Linda S. Thomashow and David M. Weller. 2001. A rapid polymerase chain reaction based assay characterizing rhizosphere populations of 2, 4-Diacetyl phloroglucinol – producing bacteria. *Phytopathology* **91**(1) : 44–54.

Buhariwalla HK, Srilaksmi P, Kannan S, Kanchi RS Chandra S, Satyaprasad K, Waliyar F, Thakur RP and Crouch JH. 2005. AFLP analysis of *Trichoderma* spp. from India compared with sequence and morphological –based diagnostics. *P Zeit* **153:** 389-400.

Bull CT, Weller DM. and Thomashow LS. 1991. Relationship between root colonization and suppression of *Gaeumannomyces graminis* var. *tritici* by *Pseudomonas fluorescens* strain 2-79. *Phytopathology* **81**: 954-959.

Burelle NK, Backman PA, Rodriguez Kabana R. and Popler LD. 1992. Potential for biological control of early leaf spot of peanut using *Bacillus cereus* and chitinase foliar amendments. *Biological Control* **2**: 321-328.

Callan NW, Mathre DE and Miller JB. 1990. Bio-priming seed treatment for biological control of *Pythium ultimum* pre—emergence damping off in Sh2 sweet corn. *Plant Disease* **74:** 368-372.

Campbell R. (1989) Biological control of Microbial Plant Pathogens. Cambridge university press, Cambridge, 218pp.

Chahal PPK and Chahal VPS. 1988. Biological control of root-knot nematode on brinjal (*Solanum melongena* L.) with *Azatobacter chroococcum*. In: Advances in Plant Nematology. Maqbool MA, Golden A, Gaffar MA and Krusberg L (eds.) National Nematological Research Centre, University of Karachi, Karachi, Pakistan, pp.257-263.

Chakraborty U and Purkayastha RP. 1984. Role of rhizobitoxine inn protecting soybean roots from *Macrophomina phaseolina* infection. *Canadian Journal of Microbiology* **30:** 285-289.

Chakraborty V, Sarkar B and Chakraborty BN. 2003. Protection of soybean root rot by *Bradyrhizobium japonicum* and *Trichoderma harzianum*, associated changes in enzyme activities and phytoalexin production. *J.Mycol.Pl.Pathol.* **33:** 21-25.

Chao WL, Nelson EB, Harman GE and Hoch HC. 1986. Colonization of the rhizosphere by biological control agents applied to seeds. *Phytopathology* **76:** 60-65.

Charudattan R. 1986. Integrated control of waterhyacinth with a pathogen, insects and herbicides. *Weed Science* **34:** 26-30.

Charudattan R. 1990c. Prospects for biological control of weeds by plant pathogens. *Fitopatol Bras*. **15:**13-19.

Charudattan R and Conway KE. 1976. *Mycoleptodiscus terrestris* leaf spot on waterhyacinth. *Plant Disease Reporter* **60:** 77-80.

Charudattan R, Linda SB, Kluepfel M and Osman YA. 1985. Biocontrol efficacy of *Cercospora rodmanii* on waterhyacinth. *Phytopathology* **75**: 1263-1269.

Chaudhary KCB and Gupta S. 1970. Control of guar caused by *Sclerotium rolfsii* with *Streptomyces nigrifaciens*. *Netherlands Journal of Plant Pathology* **76**: 326-328.

Chhabra ML, Bhatnagar MK and Sharma MP. 1992. Influence of a vesicular – arbuscular (VA) mycorrhizal fungus on important diseases of maize. *Indian Phytopathology* **54**: 235-236.

Chet, I. (1987) Innovative approaches to plant disease control John wiley and Sons New York, 372pp.

Chet, I. Ed.(1993). Biotechnology in plant disease control. Wily-Liss . New York, p.373.

Cherif M and Benhamou N. 1990. Cytochemical aspects of chitin breakdown during the parasitic action of a *Trichoderma* sp. on *Fusarium oxysporum* f.sp. *radicis – lycopersici*. *Phytopathology* **80**: 1406-1414.

Conway KE. 1976a *Cercospora rodmanii*, a new pathogen of waterhyacinth with biological control potential. *Canadian Journal of Botany* **54:**1074-1083.

Cook R J. (1990). Twenty five years of progress towards biological control in: Biological control of Soil-borne Plant Pathogens. (Ed. by D. Hornby) pp. 1-14, CAB international, Wallingford, Oxford.

Cook R J. and Baker KF. (1983). The nature and practice of Biological control of Plant Pathogens. *American Phytopathological Society*. St.Paul, Minnesota, 539.

Cook RJ and Rovira AD. 1976. The role of bacteria in the biological control of *Gaeumannomyces graminis* by suppressive soils. *Soil Biology and Biochemistry* **8**: 569-571.

David Sugar and Robert A. Spotts .1999. Control of Post harvest Decay in Pear by Four Laboratory-Grown Yeasts and Two Registered Biocontrol Products. *Plant Disease* 155–157.

De RK and Mukhopadhyay AN. 1994. Biological control of tomato damping-off by *Gliocladium virens. Journal of Biological Control* **8:** 34-40.

De la Cruz, R.E. Lorilla EB. and Aggangan, N.S. 1990. Growth of *Acacia auriculiformis* and *A. mangium* in a marginal grassland in response to inoculations with VA mycorrhiza and/or *Rhizobium.* In: Fast growing trees and Nitrogen Fixing Trees, Gusta Fischer, Stuttgart, p.321.

Desjardins and Gardner H W. 1991. Virulence of *Gibberella pulicaris* on potato tubers and its relationship to a gene for rishitin metabolism. *Phytopathology.* **81:** 429-435.

Dowling DN and O'Gara F. 1994. Metabolites of *Pseudomonas* involved in the biocontrol of plant disease. TIBTECH 12: 133-141.

Droby S, Vinokar V, Weis .B, Coihen L, Daus.A, Gold Schmidt. E, and Porat R. Induction of resistance to *Penicillium digitatum* in grape fruit by the yeast biocontrol agent *Candida oleophila. Phytopathology.* **92**: 393-399.

Druzhinina IS, Schmoll M, Seoboth B and Kubicek CP. 2006. Global carbon utilization profiles of wild type strains, mutants and transformants of *Hypocrea jecorina. Appl Environ Microbiol* **72**:2126-2133.

Dwivedi R. 1984. Biocontrol of Fusarium wilt by *Trichoderma harzianum* Rifai. *Indian Journal of Agricultural Science* **54:** 513-514.

Eapen SJ and Venugopal MN.1995. Field evaluation of *Paecilomyces lilacinus* and *Trichoderma* spp. in cardamom nurseries for the control of root-knot nematodes and rhizome rot disease. National Symposium on Nematode Problems of India- An appraisal of the Nematode Management with Eco-friendly Approaches and Bio-components. Indian Agricultural Research Institute, New Delhi. p. 43.

Edo Chalutz and Charles L. Wilson. 1990. Post harvest Biocontrol of Green and Blue Mold and Sour Rot of Citrus Fruit by *Debaromyces hansenii. Plant Disease* **74** (2) : 134–137.

Elad Y and Freeman S. 2002. Biological control of fungal plant pathogens. In: The Mycota A Comprehensive Treatise on fungi as experimental systems for basic and applied research. K Esser and JW Benett, eds. Vol. XI: Springer press, Germany Pp93-109.

Elad Y and Krishner B. 1993. Survival in the phylloplane of an introduced biocontrol agent (*Trichoderma harzianum*) and populations of the plant pathogen *Botrytis cinerea* as modified by abiotic conditions. *Phytoparasitica* **21** (4): 303-313.

Elad Y , Barak R and Henis Y. 1983. Ultrastructural studies of the interaction between *Trichoderma* spp. and plant pathogenic fungi. *Phytopathol Z* **107:** 168-175.

Elad Y, Zimand G, Zaqs Y, Zurial S and Chet I. 1993. Use of *Trichoderma harzianum* in combination or alternation with fungicides to control cucumber grey mold (*Botrytis cinerea)* under commercial green house conditions. *Pl. Pathol* **42**: 324-332.

Fan Qing and Tian Shiping. 2000. Post harvest Biological Control of Rhizopus Rot of Nectarine Fruits by *Pichia membranefaciens. Plant Disease* 1212–1215.

Garabedian S and Van Gundy SD.1983. Use of avermectins for the control of Meloidogyne incognita on tomatoes. *Journal of Nematology* **15** : 503-511.

Gerdemann JW. 1975. Vesicular-arbuscular mycorrhizae. In: The development and function of roots, Academic press, New York, pp.576-591.

Gianinazzi S. and Gianinazzi Pearson V. 1992. Cytology, histochemical and immuno-cytochemistry as tools for studying structure and function in endomycorrhiza *Methods in Microbiology.* **24:** 109-139.

Gianinazzi Pearson V., Gollette A., Dumas Gaudot E., Franken P., and Gianinazzi S. 1994. Gene expression and molecular modifications associated with plant responses to infection by arbuscular mycorrhizal fungi. In: Advances in molecular genetics of plant-microbe interactions, Kluwer , Dordsecht, pp. 179-186.

Gianinazzi Pearson V and Gianinazzi S. 1995. Proteins and protein activities in endomycorrhizal symbiosis. In: Mycorrhiza – structure, function, molecular biology and biotechnology, Springer, Heidelberg, pp. 251-266.

Gilly Zimand, Lea Valinsky, Elad Y, Chet I and Shulami Manulis. 1994. Use of the RAPD procedure for the identification of *Trichoderma* strains. *Mycological Research* **98** (5) : 531–534.

Gintis BO, Morgan- Jones G abd Rodriguez-Kabana R. 1983. Fungi associated with several developmental stages of *Heterodera glycines* from an Alabama Soybean field soil. *Nematropica* **13** : 181-200.

Grondona I, Hermosa R, Tejada M, Gomis MD, Mateos PF, Bridge PD, Monte E and Garcia-Acha I. 1997. Physiological and biochemical characterization of *Trichoderma harzianum*, a biological control agent against soil-borne fungal plant pathogens. *Applied and Environmental Microbiology* **63**(8) : 3189–3198.

Hanson LE and Howell CR. 2004. Elicitors of plant defense responses from biocontrol strains of *Trichoderma virens*. *Phytopathology* **94**:171-176.

Harrison M J. and Dixon RA. 1993. Isoflavonoid accumulation and expression of defense gene transcripts during the establishment of vesicular-arbuscular mycorrhizal associations in roots of *Medicago trancatula. Molecular plant Microbe Interactions*, **6:** 643-654.

Harman GE. 2000. Myths and dogmas of biocontrol: Changes in perception derived from research on *Trichoderma harzianum* T-22. *Pl. Dis* **84:** 377-391.

Harman GE. 2001. Microbial tools to improve crop performance and profitability and to control plant diseases. Pp 7-84. In: Int Symp Biol Contl Pl Dis New CountryMde action Application Technol. Taichun City, Taiwan.

Harman GE, Hayes CK and Ondik KL. 1998. Asexual genetics in *Trichoderma* and *Gliocladium*: Mechanisms and implications. Pp 243-270. In Kubicek CP and Harman GE, eds. *Trichoderma* and *Gliocladium.* Vol. I. Taylor & Francis, London.

Harman GE, Howell CR, Viterbo A, Chet I and Lorito M. 2004. *Trichoderma* species-opportunistic, avirulent plant symbionts. *Nat Rev Microbiol* **2** (1): 43-56.

Haware MP, Lenne JM and Mukherjee PK. 1996. Integrated management of *Botrytis* grey mold of chickpea. Proceedings of XI. International *Botrytis* Symposium 23-27 June, Wageningen p.98.

Hermosa M R, Grondona I, Iturriaga E A, Diaz-Minguez J M, Castro C, Monte E and Garcia-Acha I. 2000. Molecular characterization and identification of biocontrol isolates of *Trichoderma* spp. *Applied and Environmental Microbiology*, 1890–1898.

Hornby D.(1990). Biological control of Soil- Borne Plant Pathogens. CAB International Wallingford, Oxford, 479pp.

Howell CR. 2003. Mechanisms employed by *Trichoderma* species in the biological control of plant diseases: The history and evolution of current concepts. *Pl Dis* **87** (1): 4-10.

Howell CR. and Stipanovic RD. 1979. Suppression of *Pythium ultimum* induced damping off of cotton seedlings by *Pseudomonas fluorescens* and its antibiotic pyoluteorin. *Phytopathology* **70**: 712-715.

Howie WJ and Suslow TV. 1991. Role of antibiotic biosynthesis in the inhibition of *Pythium ultimum* in the cotton spermosphere and rhizosphere by *Pseudomonas fluorescens. Molecular Plant Microbe Interactions* 4: 393-399.

Jain RK and Sethi CL. 1988a. Influence of Endomycorrhizal fungi *Glomus fasciculatum* and *G. epigaeus* on penetration and development of *Heterodera cajani* on cowpea. *Indian Journal of Nematology* **18** : 89-93.

Jain RK and Sethi CL. 1988b. Interaction between vesicular arbuscular mycorrhiza and *Meloidogyne incognita, Heterodera cajani* on cowpea as influenced by time of inoculation. *Indian Journal of Nematology* **18 :** 263-268.

Janisieweiz WJ and Roitman J. 1988. Biological control of blue mold and grey mold on apple and pear with *Pseudomonas cepacia. Phytopathology* **78:** 1697-1700.

Jatala P, Kaltenbach R and Bocangel M. 1979. Biological control of *Meloidogyne incognita acrita* and *Globodera rostochinensis* on potatoes. *Journal of Nematology* **11:** 303.

Jesse M. Benbow and David Sugar. 1999. Fruit surface colonization and biological control of post harvest disease of pear by preharvest yeast applications. *Plant Disease* **83**(9) : 839–843.

Jeyarajan R and Angappan K. 1998. Biological suppression of plant diseases, phytoparasitic nematodes and weeds. P.48-56.

Johnson KB.1994. Dose response relationships and inundative biological control. *Phytopathology* **84:** 780-784.

Joseph Kuc. 1995. Phytoalexins, stress metabolism and disease resistance in plants. *Annu. Rev. Phytopathology*. 33:275-297.

Kajal Kumar Biswas and Anupam Varma. 2000. Identification of variants of Mungbean yellow mosaic Gemini virus by host reaction and nucleic acid spot hybridization. *Indian Phytopathology* **53** (2) : 134-141.

Kamala Nayar. 1998. Development and evaluation of a biopesticide formulation for control of foliar pathogens of rice. Ph.D. thesis, Tamil Nadu Agriculturl University, Coimbatore. pp. 198.

Kamala Nayar. And Vidhyasekaran P. 1998. Development of a powder formulation of *Pseudomonas fluorescens* PI for management of rice bacterial blight. *Annals of Applied Biology.*

Karpagam S. 1995. Variation in cultural morphology and biocontrol efficacy among protoplast regenerated isolates of *Trichoderma harzianum. Indian Journal of Mycology and Plant Pathology* **25**:102.

Keel C, Schnider U, Maurhofer M, Voisard C, Ilaville J, Burger U, Wirthner P, Haas D and Defago G. 1992. Suppression of root diseases by *Pseudomonas fluorescens* CHAO: importance of the bacterial secondary metabolite 2,4- diacetyl phloroglucinol. *Molecular Plant Microbe Interactions* **5:** 4-13.

Kempe J and Sequeira L. 1983. Biological control of bacterial wilt of potatoes: Attempts to induce resistance by treating tubers with bacteria. *Plant Disease* **67:** 499-503.

Kloepper JW. 1993. Plant growth promoting rhizobacteria as biological agents. In; Soil Microbiological Techniques. Metting B.(ed.) , Marcel Dekker Inc. New York, pp.255-274.

Kloepper JW and Schroth Mn. 1981. Relationship of *in vitro* antibiosis of plant growth promoting rhizobacteria to plant growth and displacement of root microflora. *Phytopathology* **71**: 1020-1024.

Kloepper JW, Tuzun S, Liu L and Wei G. q1993. Plant growth promoting rhizobacteria as inducers of systemic resistance. In: Pest Management : Biologically based Technologies. Lumsden RD and Waughn JL (eds.), ACS Conference Proceedings series. American Chemical Society press, pp.156-165.

Knudsen GR and Spurr HW.1987. Field persistence and efficacy of five bacterial preparations for control of peanut leaf spot. *Plant Disease* **71:** 442-445.

Koshy PK and Gulsarbanu J. 1998. Use of fungal and bacterial antagonist for the biological control of nematodes in plantation crops. Biological suppression of plant diseases, phytoparasitic nematodes and weeds p. 154-163.

Lambais, M.R. and Mehdy, M.C. 1993. Suppression of endochitinase, b-1,3- glucanases and chalcone isomerase expression in bean vesicular-arbuscular mycorrhizal roots under different soil phosphate conditions. *Molecular Plant Microbe Interactions.*, **6:** 75-83.

Lambert, D.H. and Cole, Jr.H. 1980. Effects of mycorrhizae on establishment and performance of forage species in mine spoil. *Agronomy Journal* **72**: 257-260.

Latha J, Aprajita Verma and Prasun K.Mukherjee. 2002. PCR-fingerprinting of some *Trichoderma* isolates from two Indian type culture collections – a need for re-identification of these economically important fungi. *Current Science* **83**.(4) 372–374.

Leonardo Schena, Mariella Finetti Sialer and Donato Gallitelli. 2002. Molecular detection of strain L-47 of *Aureobasidium pullalans,* a biocontrol agent of post harvest diseases. *Plant Disease* **86 :** (1) 54–60.

Lewis, J.A. and Papavizas, G.C.1991. Biocontrol of plant diseases; the approach for tomorrow. *Crop Prot.* **10:** 95-105.

Lieckfieldt E, Gary J Samuels, Helgard I.Nirenberg and Orlando Petrini. 1999. A morphological and molecular perspective of *Trichoderma viride* : Is it one or two species ? *Applied and Environmental Microbiology* **65**(6) : 2418–2428.

Lifshitz R, Kloepper JW, Kozlowski M, Simonson C, Cavison J, Tipping EM and Zaleska I. 1987. Growth promotion of canola (rape seed) seedlings by a strain of *Pseudomonas putida* under bnotobiotic conditions. . *Canadian Journal of Microbiology* **33:** 39-395.

Lih-ren Kong, Dean D.Tzeng and Chang-Hsien Yang 2001 Generation of PCR-based DNA fragments for specific detection of *Streptomyces saraceticus* N_{45}. *Proc. Natl. Sci. Counc.* Roc (B) **25**(2) : 119–127.

Linderman, R.G. 1994. Role of VAM fungi in biocontrol. In: *Mycorrhizae and Plant Health,* APS,St. Paul, pp. 1-26.

Liu L, Kloepper JW and Tuzun S. 1992. Induction of systemic resistance against cucumber mosaic virus by seed inoculation with selected rhizobacterial strains. *Phytopathology* **82:** 1109.

Liu L, Kloepper JW and Tuzun S. 1995. Induction of systemic resistance in cucumber against *Fusarium* wilt by plant growth promoting rhizobacteria. *Phytopathology* **85:** 695-698.

Mada, R.J. and Bagyaraj, D.J. 1993. Root exudation from *Leucaena leucocephala* in relation to mycorrhizal colonization. *World Journals of Microbiology and Biotechnology,* **9:** 342-344.

Maheswari TU and Mani A.1988. Combined efficacy of *Pasteuria penetrans* and *Paecilomyces lilacinus* on the biocontrol of *Meloidogyne javanica* on tomato. *International Nematology Network Newsletter* **5** : 10-11.

Maheswari TU , Mani A and Rao PK 1987. Combined efficacy of the bacterial spore parasite *Pasteuria penetrans* and nematicides in the control of *Meloidogyne javanica* on tomato. *Journal of Biological control* **1:** 53-57.

Mani A. 1988. Studies on the bacterial parasite *Pasteuria penetrans* :I. Spore viability after storage II. Culture on citrus nematode *Tylenchulus semipenetrans*. *International Nematology Network Newsletter* **5:** 24-25.

Maiti D and Sen C. 1998. Integrated biological control of *Sclerotium rolfsii* with nitrogenous fertilizers and *Trichoderma harzianum*. *Indian Journal of agricultural science* **55:** 464-468.

Mankau R. 1975. *Bacillus penetrans* causing a virulent disease of plant parasitic nematodes. *Journal of Invertebrate pathology* **26**: 333-339.

Mankau R. 1980. Biocontrol: Fungi as nematode control agents. *Journal of Nematology* **12**: 244-252.

M'Piga P, Belanger RR, Paulitz TC and Benhamou N. 1997. Increased resistance to *Fusarium oxysporum* f.sp. *radicis lycopersici* in tomato plants treated with the endophytic bacterium *Pseudomonas fluorescens* strain 6328. *Physiology and Molecular Plant Pathology* **50:** 301-320.

Mark A Roberts and Don L. Crawford. 2000. Use of Randomly Amplified Polymorphic DNA as a means of developing Genus and strain specific *Streptomyces* DNA probes. *Applied and Environmental Microbiology* **66** (6) : 2555–2564.

Mishra DK and Narain A.1994. *Gliocladium virens* and *Streptoverticillium* as sources of biocontrol of few phytopathogenic fungi. *Indian Phytopathology* **47:** 236-240.

Morandi DBJA. and Gianinazzi- Pearson, V. 1984. Isoflavonoid accumulation in soybean roots infected with vesicular arbuscular mycorrhizal fungi. *Physiology and plant pathology,* **24:** 357-364.

Morandi. D. 1996. Occurrence of phytoalexins and phenolic compounds in endomycorrhizal interactions and their potential role in biological control. *Plant and Soil* **185**: 241-251.

Morgan-Jones G, Goody G and Rodriguez-Kabana R. 1981. *Verticillium chlamydosporium*, fungal parasite of *Meloidogyne arenaria* females. *Nematropica* **11:** 115-119.

Morgan-Jones G. and Rodriguez-Kabana R. 1985. Phytonematode Pathology: Fungal modes of action- A perspective. *Nematropica* **15:** 107-114.

Morgan-Jones G, White JF. and Rodriguez-Kabana R. 1983. Phytonematode Pathology: Ultrastructural studies I-Parasitism of *Meloidogyne arenaria* eggs by *Verticillium chlamydosporium. Nematropica* **13** : 245-260.

Morgan-Jones G, White JF. and Rodriguez-Kabana R. 1984. Phytonematode Pathology: Ultrastructural studies 2-Parasitism of *Meloidogyne arenaria* eggs by *Paecilomyces lilacinus. Nematropica* **14** : 57-71.

Mrinalini C. 1995. Implementation of protoplast fusants of *Trichoderma* spp. with enhanced biocontrol efficacy to control sheath blight of rice under greenhouse conditions. *Indian Journal of Mycology and Plant Pathology* **25** :102.

Mukherjee PK. 1997. *Trichoderma* species as microbial suppressive agents of plant pathogens. In: Compendium of Agrimicrobes. Jha MN and Venkataraman GS(eds.). Today and Tomorrow Publishers. New Delhi.

Mukherjee PK and Haware MP. 1993. Biological control of *Botrytis* grey mold of chickpea. *International Chickpea Newsletter* **28:** 14-15.

Mukherjee PK and Mukhopadhyay AN. 1993. Induction of stable mutants of *Gliocladium virens* by gamma irradiation. *Indian Phytopathology* **46** : 393-397.

Mukherjee PK and Mukhopadhyay AN. 1996. Fungi as fungicides. *International Journal of Tropical Plant Diseases* **14** : 1-17.

Mukherjee PK and Raghu K. 1997a. *Trichoderma* sp. as a microbial suppressive agent of *Sclerotium rolfsii* on vegetables. *World Journal of Microbiology and Biotechnology*.

Mukherjee PK and Raghu K. 1997b. Effect of temperature on the antagonistic and biocontrol potential of *Trichoderma* sp. on *Sclerotium rolfsii. Mycopathologia* .

Mukherjee PK, Mukhopadhyay AN, Sarmah DK and Shreshta SM. 1995b. Comparative antagonistic properties of *Gliocladium virens* and *Trichoderma harzianum* on *Sclerotium rolfsii* and *Rhizoctonia solani* its relevance to understanding the mechanisms of biocontrol. *Journal of Phytopathology* **143** : 275-279.

Mukherjee PK, Shreshta SM. and Mukhopadhyay AN. 1993. Baiting with *Sclerotiumrolfsii* for selective isolation of *Gliocladium virens* from natural soil. *Biocontrol Science and Technology* **3 :** 101-104.

Mukherjee PK, Thomas P and Raghu K. 1995c. Shelf-life enhancement of fresh ginger rhizomes at ambient temperatures by combination of gamma- irradiation biocontrol and closed polyethylene bag storage. *Annals of Applied Biology* **127** : 375-384.

Mukhopadhyay AN. 2006. Policy issues and road map for popularizing and increasing the production of fungus based antagonistic organisms for crop disease management in India. In: Current status of Biological Control of Plant Diseases using Antagonistic Organisms in India. B. Ramanujan and RJ Rabindra, eds. Technical Document, Project Directorate of Biological Control, Bangalore. Pp.157-175.

Mukhopadhyay AN. 1987. Biological control of soil-borne plant pathogens by *Trichoderma* spp. *Indian Journal of Mycology and Plant Pathology* **17** : 1-10.

Mukhopadhyay AN. 1994. Biocontrol of soil-borne fungal plant pathogens – current status, future prospects and potential limitations. *Indian Phytopathology* **47:** 119-126.

Mukhopadhyay AN. 1996. Recent innovations in plant disease control by eco-friendly biopesticides. Presidential address, Agricultural Sciences section, 83rd Indian Science Congress, Patiala.

Mukhopadhyay AN, Brahmbhatt A and Patel GJ. 1986. *Trichoderma harzianum* – a potential biocontrol agent for tobacco damping off. *Tobacco Research* **12** : 26-35.

Mukhopadhyay AN and Mukherjee PK. 1998. Biological control of plant diseases: Status in India. Biological suppression of plant diseases, phytoparasitic nematodes and weeds. P.7-20.

Mukhopadhyay AN, Shreshta SM and Mukherjee PK. 1992. Biological seed treatment for control of soil-borne plant pathogens. *FAO Plant Protection Bulletin* **40 :** 21-30.

Muthamilan M and Jeyarajan R. 1996. Integrated management of *Sclerotium* root rot of groundnut involving *Trichoderma harzianum* and carbendazim. *Indian Journal of Plant Pathology* **26 :** 204-209.

Nagesh M and Parvatha Reddy P. 1995a. Comparative efficacy of *Paecilomyces lilacinus* (Thom) Samson and *Verticillium hlamydosporium* (a Zimmerman) Viegas n combination with botanicals against *Meloidogyne incognita* infecting *Crossandra undulaefolia*L. *Journal of Biological Control* **9:** 109-112.

Nagesh M and Parvatha Reddy P. 1995b. Eco-friendly management of antagonistic potential for the control of *Meloidogyne incognita* on *Crossandra undulaefoliaL*. National symposium on nematode problems of India – An appraisal of the Nematode Management with Eco-friendly Approaches and Biocomponents. Indian Agricultural Research Institute, New Delhi, pp.127-128.

Nagesh M and Parvatha Reddy P. 1996a. Management of root-knot nematodes on carnation using antagonistic fungi and neem cake. National seminar on Integrated Crop Disease Management for Sustainable Agriculture, Aurangabad.

Nagesh M. Parvatha Reddy P. and Rao MS. 1996b. Management of *Meloidogyne incognita* on *Crossandra undulaefoliaL* using arbuscular mycorrhiza *Glomus mosseae* and oil cakes. *Mycorrhiza News* **9:** 12-14.

Nagesh M. Parvatha Reddy P. and Rao MS. 1995. Management of *Meloidogyne incognita* on *Crossandra undulaefoliaL* using *Paecilomyces lilacinus* and *Pasteuria penetrans* formulations. National Symposium on Perspectives in Eco-friendly Approaches to Plant Protection. Indian Agricultural Research Institute, New Delhi, pp.47-48.

Nagesh M. Parvatha Reddy P. and Rao MS. 1997. Integrated management of *Meloidogyne incognita* on *Crossandra undulaefoliaL* using *Paecilomyces lilacinus* in combination with plant extracts. *Nematologia Mediterranea* **25**: 3-7.

Nagee A, Mukhopadhyay P N and Kothari I L 2003 Multiple Arbitary Amplicon Profiling of *Trichoderma* species known for their potential as biocontrol agents against plant pests. *J. Mycol. Pl. Pathol.,* **33**(.2) 195–199.

Nakat RV, Acharya A, Jonathan EI, Hazarika K, Jha S, Singh US, Wasnikar AR, Subba Rao DV and Sitaramaiah K. 1995. Evaluation of *Paecilomyces lilacinus* for the control of root-knot nematodes *Meloidogyne incognita* on betelvine.

National Symposium on Perspectives in Eco-friendly Approaches to Plant Protection. Indian Agricultural Research Institute, New Delhi, pp.44.

Nakkeeran S, Ramanathan A and Angappan K. 2003. Biosynthesis of antibiotics in relation to suppression of pathogens in vegetable crops. In: Biotechnological approaches for the management of plant pathogens in export oriented horticultural crops (winter school) at Tamil Nadu Agricultural University , Coimbatore. R. Samiyappan (Course Director) Pp. 204-215.

Nakkeeran S, Marimuthu T, Chandrasekar G, Sankarlingam A and Renuka Devi. 2006. Achievements in biological control of diseases with antagonistic organisms at Tamil Nadu Agricultural University, Coimbatore. In: Current status of Biological Control of Plant Diseases using Antagonistic Organisms in India. B. Ramanujan and RJ Rabindra, eds. Technical Document, Project Directorate of Biological Control, Bangalore. Pp.350-362.

Nemec S and Meridith FI. 1981. Aminoacid content of leaves in mycorrhizal and non-mycorrhizal citrus rootstocks. *Annals of Botany* **47:** 351-358.

O'Neill TM, Elad Y, Shtienberg D and Cohen A. 1996. Control of grapevine grey mold with Trichoderma harzianum T39. *Biocontrol Science and Technology* **6:** 139-146.

Padmanabhan P and Alexander KC. 1983. Seedling rot of sugarcane. Extension publication no.5, Sugarcane Breeding Institute, Coimbatore.

Papavizas, G.C.(1985). *Trichoderma* and *Gliocladium*. Biology, ecology and potential for biocontrol *A.Rev. Phytopathol.* 23 : 23.

Papavizas, G.C. and Lewis, J.A. (1988). The use of fungi in integrated control of Plant diseases. In "Fungi in Biological control systems" (Ed.by M.N.Burge) pp.235-253. Manchester University Press, Manchester.

Parvatha Reddy P and Khan RM. 1988. Evaluation of *Paecilomyces lilacinus* for the biological control of *Rotylenchulus reniformis* infecting tomato as compared with carbofuran. *Nematologia Mediterranea* **16:** 113- 116.

Parvatha Reddy P and Khan RM. 1988. Evaluation of biocontrol agent *Paecilomyces lilacinus* and carbofuran for the management of *Rotylenchulus reniformis* infecting brinjal. *Pakistan Journal of Nematology* **1**: 221-222.

Parvatha Reddy P, Khan RM. and Rao MS. Integrated management of citrus nematode *Tylenchulus semipenetrans* using pesticides and parasitic fungus *Paecilomyces lilacinus*. *Pest Management in Horticultural Ecosystems* **2:** 61-63.

Parvatha Reddy P, Rao MS. and Nagesh M. 1995. Integrated management of the citrus nematode *Tylenchulus semipenetrans* Cobb. using VA-mycorrhiza, *Glomus fasciculatum* and oil cakes. *Pest Management in Horticultural Ecosystems* **1:** 37-41.

Parvatha Reddy P and Nagesh M. 1998. Fungal and bacterial antagonists for biological suppression o plant parasitic nematodes on horticultural crops. Biological suppression of plant diseases, phytoparasitic nematodes and weeds p. 138-153.

Paulitz TC, Anas I and Fernandes DG. 1992. Biological control of *Pythium* damping off by seed treatment with *Pseudomonas putida* : relationship with ethanol production by pea and soybean seeds. *Biocontrol Science and Technology* **2:** 193-201.

Pawar AD. 2006. Registration requirements and quality parameters for antagonistic organisms. In: In: Current status of Biological Control of Plant Diseases using Antagonistic Organisms in India. B. Ramanujan and RJ Rabindra, eds. Technical Document, Project Directorate of Biological Control, Bangalore. Pp.176-188.

Pervaiz A Abbasi, Sally A Miller, Tea Meulia, Harry A Hoitink and Jin Man Kim 1999 Precise detection and tracing of *Trichoderma hamatum* 382 in compost-amended potting mixes by using molecular markers. *Applied and Environmental Microbiology* **65**(12) : 5421–5426.

Prasad RD. 2002. Formulations and applications of *Trichoderma* species for biological control of plant diseases in India. Proceedings of the ICAR-CABI workshop on Biopesticide formulations and applications, Project Directorate of Biological Control , Bangalore. P.187-202.

Putter JG. Mac Connell FA, Preiser FA, Haidri AA, Rishich SS and Dybas RA. 1981. Avermectins : Novel class of insecticides, acaricides and nematicides from a soil microorganism. *Experimentia* **37** : 963-964.

Raajimakers JM, Leeman M, Van Oorschot MMP, Van der Sluis I, Schippers B and Bakker PAHM. 1995. Dose response relationship in biological control of Fusarium wilt of radish by *Pseudomonas* sp. *Phytopathology*. **85:** 1075-1081.

Rabindran R and Vidhyasekaran P. 1996. Development of a formulation of *Pseudomonas fluorescens* PfALR2 for management of rice sheath blight. *Crop Protection* 15: 75-721.

Raghuchander T, Samiyappan R and Arjunan G. 1993. Biocontrol of *Macrophomina* root rot of mung bean. *Indian Phytopathology* **46 :** 379-382.

Ramesh Kumar N, Thirumalai Arasu V and Gunasekaran P 2002 Genotyping of antifungal plant growth – promoting Rhizobacteria, *Pseudomonas fluorescens. Current Science* **82** (12) : 1463–1466.

Rao MS. Gowen SR. Prembroke B and Parvatha Reddy P. 1996. Relationship of *Pasteuria penetrans* spore encumberance on juveniles of *Meloidogyne incognita* and their infection in adults. *Nematologia Mediterranea* **25:** 129-131.

Rao MS. Kerry BR, Gowen SR, Bourne JM and Parvatha Reddy P. 1997a. Management of *Meloidogyne incognita* in tomato nurseries by integration of *Glomus deserticola* with *Verticillium chlamydosporium. Journal of Plant Disease Protection* **104**: 419-422.

Rao MS. Mohan Das S. and Parvatha Reddy P. 1993a. Integrated management of root-knot nematode on egg plant in nursery beds with the combination of *Glomus mosseae* and neem leaf. *Indian Journal of Nematology* **23:** 14.

Rao MS. Parvatha Reddy P. and Nagesh M. 1997a. Use of biotic agent *Paecilomyces lilacinus* in combination with neem cake suspension treatment of okra seed for the management of root-knot nematode under field conditions. *National Symposium on Emerging Trends in Pest Management,* Nauni, pp.121-122.

Rao MS. Parvatha Reddy P. and Nagesh M. 1997b. Integrated management of *Meloidogyne incognita* on okra by castor cake suspension and *Paecilomyces lilacinus. Nematologia Mediterranea* **25** : 17-19.

Rao MS. Parvatha Reddy P. Somasekhar N. and Nagesh M. 1997e. Management of root-knot nematode *Meloidogyne incognita* in tomato nursery by integration of endomycorrhiza, *Glomus fasciculatum* with castor cake. *Pest Management in Horticultural Ecosystems* **3:** 31-35.

Reddy K R N, Anitha Choudary D, Anand C and Venkateswarlu P. 2004 .Genotyping of antifungal compounds producing *Pseudomonas fluorescens* strains. *J. Mycol. Pl. Pathol.,* 34: (.3) 902–905.

Richard W.Jones and Dov Prusky 2001 Expression of an antifungal peptide in *Saccharomyces* : A new approach for biological control of the post harvest disease caused by *Colletotrichum coccodes. Phytopathology,* **92** (1) : 33–37.

Rubio M B, Hermosa M R, Keck E and Monte F. 2003. Specific PCR Assays for the detection and quantification of DNA from the biocontrol strain *Trichoderma harzianum* 2413 in soil. *Microbial Ecology* **49** 25–33.

Sabitha Rani A and Reddy. G.M. 1999. Isolation of phytoalexins from groundnut leaves produced in response to infection with *Cercospora arachidicola. J.Mycol.Pl.Pathol.* **29:** 100-102.

Sailaja Rani J. and Reddi Kumar M. 2007. Protein and isozyme variability among isolates of *Pythium aphanidermatum.* J Mycol Pl Pathol 37 (1): 98-100.

Saini LC and Parashar RD. 1991. Evaluation of phylloplane microflora of cluster bean against *Xanthomonas campestris* pv. *cyamopsidis. Plant Disease Research* **6**: 120-124.

Samiyappan R. 2002. Use of formulations of bacterial antagonists in disease management in India. Proceedings of the ICAR-CABI workshop on Biopesticide formulations and applications, Project Directorate of Biological Control , Bangalore. P.180-186.

Satoshi Yamamoto and Shigeaki Harayama. 1995. PCR amplification and direct sequencing of *gyrB* genes with universal primers and their application to the detection and taxonomic analysis of *Pseudomonas putida* strains. *Applied and Environmental Microbiology,* 1104–1109.

Sawant IS and Mukhopadhyay AN. 1990. Integration of metalaxyl with *Trichoderma harzianum* for the control of *Pythium* damping off in sugarbeet. *Indian Phytopathology* **43:** 535-541.

Sawant IS and Sawant SD. 1996. A simple method for achieving high C.F.U. of *Trichoderma harzianum* on organic wastes for field applications. *Indian Phytopathology* **49:** 185-187.

Schonbeck, F. 1979. Endomycorrhiza in relation to plant diseases. In: Soil-borne plant pathogens. Academic Press, New York. pp. 271-280.

Schenck, N.C. 1987. Vesicular- arbuscular mycorrhizal fungi and the control of fungal root disease. In: Innovative approaches to plant disease control. Wiley, New York, pp.179-191.

Schneider, R.W., Ed. (1982). Suppressive soils. American Phyto-pathological society, st. paul, Minnesota, 88pp.

Secilia, J. and Bagyaraj, D.J. 1987. Bacteria and actinomycetes associated with pot cultures of vesicular arbuscular mycorrhizas. *Canadian Journal of Microbiology*, **33:** 1069-1073.

Shekhawat PS and Chakrabarti BP. 1977. Epiphytic bacterial and actinomycetes population and control of bacterial leaf spot of chillies by an antagonistic bacterium associated with bacterial lesions. *Indian Phytopathology* **30:** 358-360.

Singh A and Singh HB. 2004. Control of collar rot of mints (*Mentha* spp.) caused by *Sclerotium rolfsii* by using biological means. *Curr sci* **87** (3): 362-366.

Singh A, Srivastava S. and Singh HB. 2007. Effect of substrates on growth and shelf life of *Trichoderma harzianum* and its use in biocontrol of diseases. *Bio-resources Technol.* **98:** 470-473.

Singh HB. 2006. Achievements in biological control of diseases with antagonistic organisms at National Botanical Research Institute, Lucknow. In: Current status of Biological Control of Plant Diseases using Antagonistic Organisms in India. B. Ramanujan and RJ Rabindra, eds. Technical Document, Project Directorate of Biological Control, Bangalore. Pp.329-340.

Singh HB, Karla A and Patra NK. 2001. Potential for biological control of wilt and rot in mints by Trichoderma harzianum and Gliocladium virens. *Kavaka* **28 & 29** : 45-48.

Singh, HB 2006. Trichoderma: A Boon for Biopesticide Industry: Prof. Bhide Memorial Award Lecture. *J Mycol Pl Pathol* **36** (3) 373-384.

Singh US and Najam Waris Zaidi 2002. Current status of formulation and delivery of fungal and bacterial antagonists for disease management in India. Proceedings of the ICAR-CABI workshop on Biopesticide formulations and applications, Project Directorate of Biological Control , Bangalore. P.168-179.

Sikora RA. 1979. Predisposition to *Meloidogyne* infection by the endotrophic mycorrhizal fungus Glomus mosseae. In Root-knot Nematodes Systematics Biology and Control. Lamberti F and Taylor CE (eds.) academic Press, London, pp. 399-404.

Sitaramaiah K. and Sikora RA. 1982. Effect of mycorrhizal fungus *Glomus fasciculatum* on the host –parasite relationship of *Rotylenchulus reniformis* in tomato. *Nematologica* **28:** 412-419.

Sivaramakrishnan S, Seetha Kanan and Singh S D 2002 Detection of genetic variability in *Fusarium udum* using DNA markers. *Indian Phytopathology* **55:** 258-263.

Sneh EW, Dupler M, Elad Y and baker R.1984. Chlamydospore germination of *Fusarium oxysporum* f.sp. *cucumerinum* as affected by fluorescent and lytic bacteria from fusarium suppressive soil. *Phytopathology* **74:**1115-1121.

Sosamma VK. And Koshy PK. 1995. Effect of *Pasteuria penetrans* and *Paecilomyces lilacinus* on population build up of root-knot nematode *Meloidogyne incognita* on black pepper. National Symposium on Nematode Problems of Indi- An Appraisal of the Nematode Management with Eco-friendly Approaches and Biocomponents. Indian Agricultural Research Institute, New Delhi, p. 47.

Sreenvasaprasad S and Manibhushanarao K. 1990. Biocontrol potential of fungal antagonists *Gliocladium virens* and *Trichoderma longibrachatum*. *Journal of Plant Disease Protection* **97**: 570-579.

Sreenvasaprasad S and Manibhushanarao K. 1993. Efficacy of *Gliocladium virens* and *Trichoderma longibrachiatum* as biological control agents of groundnut root and stem rot diseases. *International Journal of Pest Mnagement* **39 :** 167-171.

Stutz EW, Defgo G and Kern H. 1986. Naturally occuring fluorescent pseudomnads involved in suppression of black root rot of tobcco. *Phytopathology* **76**: 181-185.

Suslow TV and Schroth MN. 1982. Rhizobacteria of sugarbeets: effects of seed application and root colonization on yield. *Phytopathology* **72**: 199-206.

Tegtmeier. K.J. and Van Etten. H.D. 1982. The role of pisatin tolerance and degradation in the virulence of *Nectria haematococca* on peas: A Genetic Analysis. *Phytopathology.* **72:** 608-612.

Templeton GE. 1982. Status of weed control with plant pathogens. In: Biological control of weeds with plant pathogens. Charudattan R and Walker HL (eds.) pp.29-44. John Wiley, New York.

Templeton GE, TeBeest DO and Smith RJ Jr. 1979. Biological weed control with mycoherbicides. *Annual Review of Phytopathology* **17:** 301-310.

Tousaan, T.A. (1975). *Fusarium.* Suppressive soils. In Biology and control of soil-borne plant pathogens (Ed.by G.W.Bruchi) pp. 145-151. American Phytopathological Society St.Paul, Minnesota.

Utkhede RS and Rake JE. 1980. Biological control of onion white rot. *Soil Biology and Biochemistry* **12:** 101-105.

Van Peer. R, Niemann. G.J and Schippers. B. 1991. Induced resistance and phytoalexin accumulation in biological control of *Fusarium* wilt of carnation by *Pseudomonas* sp. Strain WCS filter. *Phytopathology* **81**: 728-733.

Van Peer R. and Schippers. B. 1992. Lipopolysaccharides of plant growth promoting *Pseudomonas* sp. strain WCS 417r induce resistance in carnation to *Fusarium* wilt. *Netherlands Journal of Plant Pathology* **98**: 129-139.

Vasudeva RS and Roy TC. 1950. The effect of associated soil microflora on *Fusariumudum* Butler, the fungus causing wilt of pigeonpea. *Annals of Applied Biology* **37:** 167-177.

Vasudeva RS, Jain AC and Nema NG. 1952. Investigations of the inhibitory action of *Bacillus subtilis* on *Fusarium udum* Butler, the fungus causing wilt of pigeonpea. *Annals of Applied Biology* **39**: 229-238.

Vasudeva RS,Subbaiah TV, Sastry MLM. Rangaswamy G and Ayengar RS. 1958. Bulbiformin, an antibiotic produced by *Bacillus subtilis, Annals of Applied Biology* **46**: 336-345.

Vasudeva RS, Singh GP and Iyengar MRS. 1962. Biological activity of bulbiformin in soil. *Annals of Applied Biology* **50** : 113-119.

Vasudeva RS, Singh P, Sengupta PK and Mahmood M. 1963. Further studies on the biological activity of bulbiformin. *Annals of Applied Biology* **51** : 415-423.

Vast, P. 1997. Interaction of endomycorrhiza of Arabica coffee and the chemical control of nematodes (*Pratylenchus coffeae* and *Meloidogyne konaensis*). Dix- septieme colloque Scientifique International sur le Café, Nairobi, Kenys, 20-25 juillet, pp.564-571.

Venkateswarlu R and Reddi Kumar M. 2007. Studies on genetic variability among isolates of *Trichoderma* spp. by RAPD and ITS-PCR. J Mycol Pl Path. (sent for publication)

Vey A, Hoagland RE and Butt TM. 2001. Toxic metabolites of fungal biocontrol agents. Fungi as biocontrol agents: Progress, Problems and Potential. Butt TM, Jackson C and Magan N, eds. Pp.311-346. CAB International, Bristol.

Vidya Sekharan. P. 1997. Fungal pathogenesis in plants and crops: Molecular Biology and Host Defence Mechanisms. Marcel Dekker, New York,pp.553.

Vidya Sekharan. P. 1998. Biological suppression of major diseases of field crops using bacterial antagonists. Biological suppression of plant diseases, phytoparasitic nematodes and weeds p. 81-95.

Vidya Sekharan. P. and Muthamilan M. 1995. Development of formulations of *Pseudomonas fluorescens* for control of chickpea wilt. *Plant Disease* **79**: 782-786.

Vidya Sekharan. P. and Muthamilan M. 1998a. Evaluation of a powder formulation of *Pseudomonas fluorescens* Pf1 for controlling rice sheath blight. *Biocontrol Science and Technology.*

Viterbo A, Harrel M, Horwitz BA, Chet I and Mukherjee PK. 2005. *Trichoderma* MAP –kinase signalling is involved in induction of plant systemic resistance. *Appl Environ Microbiol* **71**: 6241-6246.

Volpin, H., Phillips, D.A., Okon, Y. and Kapulnik, Y. 1995. Suppression of an isoflavonoid phytoalexin defense response in mycorrhizal alfalfa roots. *Plant Physiology*, **108**:1449-1454.

Wajid Khan M. 1998. Present status of biological suppression of plant parasitic nematodes. Biological suppression of plant diseases, phytoparasitic nematodes and weeds p. 118-127.

Walters DR, Newton AC and Lyon GD. 2005. Induced resistance : helping plants to help themselves. Biologist 52: 28-33.

Walther D and Gindart D. 1988. Biological control of damping off of sugarbeet and cotton with *Chaetomium globosum* or a fluorescent *Pseudomonas* sp. *Canadian Journal of Microbiology* **34**: 631-637.

Weindling R.1932. *Trichoderma lignorum* as a parasite of other soil fungi. *Phytopathology* **22** : 834-845.

Weindling R. 1934. Studies on a lethal principle effective in the parasitic action of *Trichoderma lignorum* on *Rhizoctonia solani* and other soil fungi. *Phytopathology* **24** : 1153-1179.

Weindling R and Emerson OH. 1936. The isolation of a toxic substance from the culture of a *Trichoderma. Phytopathology* **26**: 1068-1070.

Weindling R and Fawcett HS. 1936. Experiment in the control of Rhizoctonia damping off of citrus seedlings. *Hilgardia 10*: 1-16.

Weller, D.M. 1988. Biological Control of Soil borne plant pathogens in the rhizosphere with bacteria. *A.Rev.Phytopathol* **26:** 379-407.

Weller MD and Cook RJ. 1983. Suppression of take-all of wheat by seed treatment with fluorescent pseudomonads. *Phytopathology* **73**: 463-469.

Whipps, J.M. 1992. Status of biological disease control in horticulture. *Biocontrol Sci. Technol.* **2**: 3-24.

Whipps JM. 1997. Development in the biological control of soil- borne plant pathogens. *Advances in Botanical Research* **26**: 1-134.

Wilson M and Lindow SE. 1993. Interaction between the biological control agent *Pseudomonas fluorescens* strain A506 *Erwinia amylovora* in pear blossoms. *Phytopathology* **83**: 117-123.

Woo S L, Donzelli B, Scala F, Mach R, Harman G E, Kubicek C P, Delsorbo G and Lorito M 1999 Disruption of the ech 42 (Endochitinase – Encoding) Gene Affects Biocontrol Activity in *Trichoderma harzianum* P1. *Molecular Plant – Microbe Interactions* **12** (5) : 419–429.

Xiao- Yan S, Qing-Tao S, Shu- Tao X, Xiu-Lan C, Cai-Yun S, and Yu- Zhong Z. 2006. Broad spectrum antimicrobial activity and high stability of Trichokonins from *Trichoderma koningii* SMF2 against plant pthogens. *FEMS Microbiol Lett* **260**:119-125.

Xu GW and Gross DC. 1986. Selection of fluorescent pseudomonads antagonistic to *Erwinia caratovora* and suppression of potato seed piece decay. *Phytopathology* **76**: 414-422.

Yasuomi Tada, Shingo Hata' Hitoshi Nakayashiki, Yukio Tosa, and Shigeyuki Mayama. Signal mediators for phytoalexin production in defense response of oats elicited by victorin as a specific elicitor. *J.Gen.Plant Pathol.* **66**: 185-190.

Yedidia I, Benhamou N, Kapulnik Y and Chet I. 2000. Induction and accumulation of pathogenesis related protein activity during the early stage of root colonization by *Trichoderma harzianum. Pl Physiol Biochem* **38**: 863-873.

Yedidia I, Shoresh M, Kerem Z, Benhamou N, Kapulnik Y and Chet I. 2003. Concomitant induction of systemic resistance to *Pseudomonas syringae* pv. *lachrymans* in cucumber by *Trichoderma asperellum* (T-203) and accumulation of phytoalexins. *Appl Environ Microbiol* **69**: 7343-7353.

Zdor RE and Anderson AJ. 1992. Influence of root colonizing bacteria on the defense responses of bean. *Plant and Soil* **140**: 99-107.

Zeillinger S, Galhaup C, Payer K, Woo SL, Mach RL, Fekete C, Lorito M and Kubicek CP. 1999. Chitinase gene expression during mycoparasitic interaction of *Trichoderma harzianum* with its host. *Fungal Gene Biol* **26**: 131-140.

Zhang J, Howell CR and Starr JL. 1996. Suppression of Fusarium colonization of cotton roots and Fusarium wilt by seed treatment with *Gliocladium virens* and *Bacillus subtilis. Bioconl Sci Technol* **6:** 175-187.

References

Ahmed, J.S. and Baker, R. 1988. Implications of Rhisphere competence of Trichoderma harzianum. Canadian Journal of Microbiology, 34, 229-234.

Annonymous, 1999 Guidelines / data requirements for registration of biopesticides. Directrate of plant protection, Quarantive and storage. Ministry of Agriculture, Government of India.

Baby, V.I. and Manibhushan Rao, K. 1996. Fungal antagonists and VA mycorrhigal fungi for Biocontrol of Rhizoctonia solani, the rice sheath blight pathogen In : Recent developments in Biocontrol of Plant Pathogens. Manibhushan Rao, K. and Mahadevan, A. (Eds.) PP. 1-9.

Baker, R. 1990. An overview of current and future strategies and models for Biological control. In : Biological control of Soil borne plant pathogens. (eds. D.Honby) AB international walling ford, united Kingdom. Pp 375-388.

Connick WJ Jr., lweis JA and Quimby PC Jr. 1990. Formulation of biocontrol agents for use in plant pathology. In : New directions in Biological control P: Alternatives for suppressing Agril. Pests and Diseases. Baker RR and Dunn. PE (Eds), A.R.Liss, New York pp. 345-372.

Dandurand, L.M., and Khundsen, G.e. 1993. Influence of Pseudomonas fluorescens on hyphal growth and biocontrol activity of Trichoderma harzianum in the spermosphere and rhizosphere of the pea. Phytopathology, 83 : 265-270.

Delacruz, J., Pinor – Toro, J.A. Benitez, T., Lobell, A., and Romero, L.C. 1995. A norel endo – b-1, 1-glucanase, BGN 13.1 involved in the mycoparasitism of Trichoderma harzianum. Journal of Bacteriology, 177 : 6937 – 6945.

Fravel Dr. 1988. Role of Antibiosis in the biocontrol of plant diseases. Annual Review of Phytopathology 26 : 75-91.

Harman, G.E. and B Jooljman, T. 1998. Potential and existing uses of Trichoderma and Gliocladium for Plant disease control and plant grouth enhancement. In : Trichoderma and Gliocladium Vol. 2

(eds. G.E. Harman and C.K. Kubicek,) Taylor and Francis Ltd., London pp. 229-265.

Jayarajan R, Ramakrishnan G, Divakaran D and Sridhar R. 1994. Development of products of Trichoderma viride and Bacillus subtitles of biocontrol of root rot diseases. In : Biotechnology In India Dived (ed.), Bioved Research society, Allahabad, pp. 25-36.

Kabra A, Singh HB, Panday R, Patra NK, Kaliyar N, Gupta ML, Dhawal OP and Kumar S. 2002. Strain of Trichoderma harzianum useful as Nematode inhibitor, fungicide and plant growth promotger and a process for the isolation these of (US patent No. 6, 475, 772).

Kousalya G and Jayarajan R. 1988. Techniques for mass multiplication of Trichoderma viride and T.harzianum proceedings of National seminar on Management of crop diseases with plant products / Biological agents , 10-12 Janchay 1988, Agricultural college and research Institute, Madurai, pp. 32-33 (Abstract).

Kumar, A. and Gupta, J.P. 2000. Variation in the enzymatic activity of tebucorazole tolerant biotypes of Trichoderma viride. In: proceedings of Indian Phytopathological society golden jubilee, International Conference on "Integrated Plant Disease Management for Sustainable Agriculture ", pp 391-392.

Leeman, M., Den ouder, F.M., Van Pelt., J.A., Dirkx, f.p.m, STerijl, H., Bakkeer, P.A. H.M., and Schippers, B. 1996. from availability affects induction of systemic Resistance to Fusarium wilt of raddish by Pseudomonas Alumorescens. Phytopathology, 86; 149-155.

Lenis JA, Papavizas GC and Lumsden RD 1991. A new formulation system for the application of biocontrol fungi to soil. Biocontrol science and Technology 1 : 59 – 69.

Limon, M.C., Lora, J.M., Garcia, T., Delacruz, T., Lobell A., Benifez, T. and Pintol – Tora, J.A. 1995. Primary structure and expression of 33 K Da Chitinase gene from mycoparasitic fungus Trichoderma harzianum. Current Genetics, 28: 478-483.

Liu SD, 1988. The use of Benomyl resistance mutant of Trichoderma Koningii as a biocontrol agent against root rot diseases of Chysanthenum and adjukibean. Trichoderma News letter, 4: 5 (abstract).

Lora, J.M., De la Curz, J., Benifez, T. and pintor Toro, J.A. 1995. Molecular characterization and heterogeous expression of an endo – β-1, 6-glucanase gene from the mycoparasitic fungus Trichoderma harzianum. Mol. Gen. Genetic., 247 : 639-6 45.

Mishra, D.S., Singh, U.S. and Durivedi 2001. Relative efficacy of conidia and chlamydospores based formulations of Trichoderma spp. In: Biopesticide conference in a emuging trends in BC : BET" held at Chandigarh, Hariyana, Feb. T-9, 2001, 81 P (Abstr).

Mukherjee, P.K., Haware, M.P. and Raghu, P. 1997. Induction and evaluation of brought-tolerant mutants of Trichoderma viride for biological control botrytis gray -mould of chickpea. Indian Phytopathology 50 : 485-489.

Mukhopadhyay, A.N. Shrestha, S.M. and Mukherjee, P.K. 1992. Biological seed treatment for control of soil-borne plant pathogens. FAO Pl. Plot. Bull., 40: 21-30.

Nakkeeran, S. 2001. Genetic improvement of Trichoderma viride and assessment of its efficacy against Soil-borne pathogens, Ph,D. Thesis, TNAU, Coimbatore, p-148.

Nakkeeran, S. and Devi, P.R. 1997. Seed borne micro borne of pigeon pea and their management. Plant Disease Research, 2 : 197-200.

Nakkeran, S., Sabitha Doraiswamy and Samiyappan, R. 2001. Genetic improvement of Trichoderma viride and arresment of its efficacy against Soil – borne pathogens. In : National Seminar and Emerging Trends in pests and Diseases and their management centre for plant protection studies, TNAU, Coimbatore P130.

Nakkerran., S. and Jayarajan, R.1996. Exploitation of antagonistic potential of Trichoderma for field use. National symposium on disease of plantation crops and their management. Institute of Agriculture, Sriniketan.

Panlitz, T.C. and Belonger, R.R. 2001. Biological control in Green house systems. Anmal Review of Phytopathology 39 : 103-135.

Prasad, R.D. and Rageswaran, R.2000 C. Effect of soil application of a gramlar formulation of Trichoderma harzianum on feed rot and damy – off of chickea incited by Rhizoctonia solani, saprophytioc grounts of the pathogen and bioagent proliferation. Journal of mycology and plant pathology 30 : 216 -220.

Prasad, R.D. and Rajeswaran, R.2000. shelf life and Bioefficacy of Trichoderma harzianum formulated in various courier materials. Plant Disease Research; 15 : 38-42.

Prasad, R.D. and Rangeswaral, R. 2000 a. An improved medium for mass production of the biocontrol fungus Trichoderma harzianum journal of mycology and plant pathology 30 : 233-235.

Ramakrishnan G, Jayarajan R and Divakaran D. 1994. Tak based formulation of Trichoderma viride for biological control of Macrophomina phaseolina. Journal of Biological control 8 : 41-44.

Rjappan, K. 1997. Induction of mutants in Trichoderma viride by UV – irradiation. Plant Disease Research 12 (1): 1-5.

Selvakumar, R., Sivastava, K.d., Aggarwal, Rashmi., Singh, D.V. and Duraja prem. 2000. Studies on development of Trichoderma viride mutants and their effect on Ustilago regetum var. tritici Indian Phytopathology 53 : 185-189.

Singh A, Srivastava S and Singh HB. 2007. Effect of substrates on growth and shelf life of Trichoderma harzianum and its use in biocontrol of diseases. Bio resource Technology, 98 : 470-473.

Tiwari, A.K. 1996. Biological control of chick pea wilt – complex using different formulations of Gliocladium virens through seed treatment. Ph.D. thesis, G.B. Plant University of Agriculture and Technology., Pantnagar, India, 167 p.

Troutman, J.C. and Matezka, J.C. 1978. Induced Tolerance of Trichoderma viride to benomyl. Phytopathological News, 12 : 131 (abstr.).

Vidya R. 1995. Studies on biological control of mycyheam root rot (Macrophomina Phaseolina (Tami) Goid. by Trichoderma viride pers. M.Sc. Thesis, TNAU, Coimbatore, 145 + Xvi P.

Zaidi N.W and Singh U.S 2004 b. Use of FYM for mass multiplication and delivery of BC agents. T.harzianum and pseudomonas fluorescens. Asian Agri History 8 : 297 – 304.

www.ingramcontent.com/pod-product-compliance
Ingram Content Group UK Ltd.
Pitfield, Milton Keynes, MK11 3LW, UK
UKHW021323190726
13845UKWH00003B/36

9 789352 300648